SpringerBriefs in Environmental Science

For further volumes:
http://www.springer.com/series/8868

David W. Hursh · Camille A. Martina
With Michael A. Trush and Hillary B. Davis

Teaching Environmental Health to Children

An Interdisciplinary Approach

Springer

David W. Hursh
Warner Graduate School of Education
University of Rochester
Rochester, NY 14627-0425
USA
e-mail: dhursh@warner.rochester.edu

and

Center for Globalization and Sustainable Development
The Earth Institute
Columbia University
New York, NY 1005
USA

Camille A. Martina
Community and Preventive Medicine
University of Rochester
School of Medicine and Dentistry
Rochester, NY 14642
USA
e-mail: camille_martina@urmc.rochester.edu

Michael A. Trush
Department of Environmental Health Science
Johns Hopkins Bloomberg School of Public Health
North Wolfe St. 615, Rm. E7537
Baltimore, MD 21205
USA
e-mail: mtrush@jhsph.edu

Hillary B. Davis
Technology for Learning Consortium, Inc.
Sauga Ave. 75
North Kingstown
RI 02852-4515
USA
e-mail: hilarie@techforlearning.org

ISSN 2191-5547
ISBN 978-94-007-1810-4
DOI 10.1007/978-94-007-1811-1
Springer Dordrecht Heidelberg London New York

e-ISSN 2191-5555
e-ISBN 978-94-007-1811-1

Cover design: eStudio Calamar, Berlin/Figueres

Printed on acid-free paper

Springer is part of Springer Science+Business Media (www.springer.com)

Preface

No matter where we live, we are surrounded by manufactured and natural substances that can negatively affect our health. In the developing world two million people die every year from causes associated with exposures to smoke from cooking with biomass and coal (Legros et al. 2009). Others become ill or die from disassembling electronic waste for the metals, like lead, contained within (Gottesfeld 2009). Some estimate that 750,000 Chinese die prematurely each year from air and water pollution (Barboza 2007). In the United States, the Environmental Protection Agency (2004) reports that 44% of streams and 64% of lakes were unfit for fishing and swimming. And, every day we are learning about new dangers such as bisphenal A, perchlorate, and fire retardants.

Based on our experience teaching about environmental health and our collaboration with the projects described in this text, we are convinced that students of all ages—from elementary through secondary school—are interested in and can benefit from learning about environmental health. Furthermore, students can play a significant role in educating other children and adults, and changing their community's practices and policies. Learning about environmental health should not be merely an academic exercise, but instead an opportunity for students to make a difference in their own and other's lives. Indeed, a global movement to teach environmental health in schools could prevent thousands of illnesses and save thousands of lives.

The lessons described in this text are based on nine Environmental Health Projects funded by the United States National Institute of Environmental Health Sciences (NIEHS). Complete descriptions of the literally hundreds of lessons created by the different projects are available on the NIEHS website (http://www.niehs.nih.gov/health/scied/teachers/curricular.cfm). Because we neither can nor should reproduce in this text all of the lessons located on the website, we have aimed, instead, to provide educators with some of the key concepts in learning about environmental health, and provided some exemplary lessons regarding air and water pollution, heavy metals, and food. We hope that educators will adapt the pedagogical approaches that we suggest to environmental health issues in their own communities. Together, we can make a difference.

References

Barboza, D. (2007, July 5). China reportedly urged omitting pollution-death estimates. *The New York Times*.

Gottesfeld, P. (2009, February 20). Lead in the developing world. *Living on Earth* Accessed December 5, 2010 at: http://www.loe.org/shows/segments.htm?programID=09-P13-00008&segmentID=8

Legros, G., Havet, I., Bruce, N. and Bonjour, S. (2009). The energy access situation in developing countries: a review focusing on the least developed countries and Sub-Saharan Africa. UNDP-WHO. Available at http://content.undp.org/go/newsroom/publications/environment-energy/www-ee-library/sustainable-energy/undp-who-report-on-energy-access-in-developing-countries-review-of-ldcs—ssas.en

United States Environmental Protection Agency (2004). *National water quality inventory: Report to congress, 2004 reporting cycle*. Washington. D.C.: US Environmental Protection Agency. http://water.epa.gov/lawsregs/guidance/cwa/305b/2004report_index.cfm

Acknowledgements

This book comes out of the work of the project Environmental Health Science as an Integrative Context for Learning (EHSIC), funded by that National Institute for Environmental Health Sciences (NEIHS). David W. Hursh and Camille A. Martina would like to thank all the grant participants, including teachers in the classrooms from the nine sites around the United States who developed and implemented the curriculum described here. We would also like especially to thank Liam O'Fallon of the NIEHS, who directed the project.

We also thank Michael A. Trush and Hillary B. Davis for their work on chapters two through four. We describe the work of two teachers, in particular, whose work we found inspiring: Mike Fantauzzo, social studies teacher in the Rush-Henrietta Central School District (New York) and Dan Sullivan, science teacher in the Rochester City School District (New York).

Several of David's former and current graduate students were helpful in reading and critiquing the manuscript, including Joe Henderson, Oliver Cashman-Brown, and Sabrina Kahn. Janet Moore, administrative assistant, assisted with the manuscript.

Dr. Trush acknowledges support from NIEHS Center Grant ES03819. He acknowledges his colleagues at Maryland Public Television, Gail Long, Elisa Hozore, and Cindy Mutryn for the over 15 years of dedication working on NIEHS-supported grants to bring environmental health to teachers and their students. More importantly, Dr. Trush would like to acknowledge and thank the many teachers he interacted with on these projects for their efforts in bringing the connection between the environment and human health to their students.

Dr. Davis acknowledge support from NIEHS for RSMAS AMBIENT project, and her colleagues at the University of Miami, Lora Fleming, Ken Goodman and Lisa Pitman for many years of outstanding work on environmental health topics with Dade and Broward county teachers and students. Dr. Davis would especially like to acknowledge the teachers who so powerfully used the ethical reasoning protocol to ignite their students' interest and passion for environmental health issues in their own communities, particularly Fred Matter and Martha Adams.

Acknowledgements

Contents

Environmental Health as an Interdisciplinary Subject

David W. Hursh and Camille A. Martina

Abstract Lessons on environmental health can be integrated throughout the school curriculum. In chapter one, we first situate the book within the conceptual constructs of the curriculum goals of the National Institute for Environmental Health Sciences (NIEHS) in the United States, which funded developing and implementing the curriculum described. Second, we provide an overview of environmental health along with some of the key themes and concepts that are introduced in the lessons, including environmental agents, toxins and toxicants, dose response, and action strategies that can be taken to reduce exposure risks. Third, we highlight a few of the literally hundreds of lessons developed. Finally, we provide examples of suggested activities in the following chapters that may be adapted for use in schools throughout the world.

Keywords Environmental agents · National Institute of Environmental Health Sciences · Interdisciplinary curriculum · Exposure risk · Problem-based learning

Introduction

Every day the media provide numerous examples of the relationship between human health and the environment, such as the danger from bisphenol-A (BPA) in plastics, pollution from chemical plants in China, and increased incidences of pulmonary and cardiac disease due to air pollution and ultra-fine volatile organic compound (VOC) particulates from motor vehicle exhaust, and, in the United States, food born illnesses that kill 5,000 people per year (Layton 2009), Hurricane Katrina flooding New Orleans with a toxic soup, and children who suffer lead poisoning from lead-based paints in older homes. In fact, we live in a world of both human produced and natural environmental hazards.

D. W. Hursh and C. A. Martina, *Teaching Environmental Health to Children*,
SpringerBriefs in Environmental Science, DOI: 10.1007/978-94-007-1811-1_1,

Given the pervasiveness and enormity of these environmental problems, it is inexcusable that schools rarely teach students to evaluate their risks and reduce their exposure to toxicants in our environment. If these topics are taught at all, they are typically limited to a course in environmental science, which is often an elective course of study. In response, we suggest that students are likely to be familiar with some of these issues and not only would like to be informed of the risks they face, but that we have a moral obligation to assist them in assessing and reducing those risks. Furthermore, once students graduate from compulsory education, they may never have another opportunity to examine these issues in a thoughtful and sustained fashion. Schools are, therefore, the best and often only opportunity to educate children about toxins (an agent from a natural source, such as molds or animal venoms) and toxic agents (such as chemicals and gases) in their environment, and how they might reduce their own, their family's and community's risk to these exposures. They will also be equipped to investigate what public health policies might be implemented to reduce their risk.

Understanding and reducing toxic health risks requires more than only a scientific explanation of the problem. Rather, teachers and students need to incorporate broader social, political, historical, philosophical and ethical questions having to do with environmental justice, and the changing relationship between people and their environment, as well as citizen action. Consequently, environmental health necessitates an interdisciplinary approach and can be incorporated into almost any grade and subject area.

Environmental health is not only important in and of itself, but can facilitate students' learning in science, math, language, history, and civics. The National Institute of Environmental Health Sciences (NIEHS) in the United States, over the last several decades, has funded university, elementary, and secondary educators to collaborate in developing and implementing integrative interdisciplinary environmental health curriculum in schools. This volume attempts to capture just some of the efforts of nine Environmental Health Project sites across the United States.

Together the project sites created literally hundreds of lessons, almost all of which are available on the NIEHS website at http://www.niehs.nih.gov/health/scied/teachers/curricular.cfm. Because we cannot reproduce all or even most of the lessons here, we have aimed, instead, to provide an overview of some of the lessons, including the rationale, methods, and resources so that educators have a basis for developing lessons appropriate for their students, schools, and communities. Most of the lessons, we suggest, can be adapted to almost any grade level and can be implemented in either a specific course on environmental health or in other subject areas. Ideally, we suggest, teachers in schools should work together to develop an environmental health curriculum that begins in elementary school and continues through commencement. As we will describe below, over the course of their educational career, students can gain an increasingly sophisticated and complex understanding of environmental health.

In this introduction, then, we do the following: First, we situate the book within the conceptual constructs of the curriculum goals of the NIEHS in the United States, which funded developing and implementing the curriculum described. Second, we provide an overview of environmental health along with some of the key themes and concepts that are introduced in the lessons, including environmental agents, toxins and toxicants, dose response, and action strategies that can be taken to reduce students' and others' risk. Third, we highlight a few of the literally hundreds of lessons developed. Finally, examples will be provided regarding how the suggested activities in the following chapters may be adapted for use in schools throughout the world. Appendix A provides the reader with abbreviated descriptions of all of the units and lessons developed through this project.

The nine Environmental Health Project Sites that were part of this NIEHS funding are listed below (web links are provided in Chap. 3, *Environmental Health Curriculum*):

- Atmospheric and marine-based interdisciplinary training (AMBIENT), University of Miami, Coral Gables, FL, USA), Grades 9–12.
- ECOS: The environment as a context for opportunities in school, Baylor College of Medicine (Houston, TX, USA), Grades K-5.'

- ENVIROHEALTH CONNECTIONS: A collaborative exploration of the environment and human health, Maryland Public TV (Owings Mills, MD, USA), Grades 6–9.
- Project EXCITE Environmental explorations through cross-disciplinary investigation team experiences. Bowling Green State University (Bowling Green, OH, USA), Grades 4–10.
- Hydroville: Learning Through Environmental Health Science Scenarios, Oregon State University (Corvallis. OR, USA), Grades 9–12.
- IEHMSP Integrated Environmental Health Middle School Project University of Washington (Seattle, Washington, USA), Grades 6–9.
- My Environment, My Health, My Choices: Problem-Based Learning (PBL) for Environmental Health. University of Rochester (Rochester, NY, USA), Grades 5–12.
- PEER Partnership for environmental education and rural health: integrating environmental health science in rural schools, Texas A&M University (College Station, TX, USA), Grades 6–9.
- SUC_2ES_2 Students Understanding Critical Connections between the Environment, Society, and Self, University of Medicine and Dentistry of New Jersey (New Brunswick, NJ, USA). Grade 2, 5, 7.

While the curriculum projects focused on different grades and age groups [grades Kindergarden-5th grade are the elementary school students ages 5–11, grades 6–8 are middle school students ages 12–14, and grades 9–12 are secondary or high schools ages 15–18, most of the units and lessons can be adopted for younger or older children than those originally intended.

The National Institute of Environmental Health Sciences

This volume comes out of the efforts of educators at nine different sites in the United States, funded by a grant from the NIEHS, to develop and implement curriculum on the relationship between the environment and human health. The goal of the NIEHS is to reduce the burden of human illness and disability by understanding how the environment influences the development and progression of human disease by focusing on basic science and disease-oriented research, global environmental health initiatives, and multidisciplinary training for researchers.

In the early 1990s, the NIEHS established the goal of developing an environmental health sciences education program from elementary through secondary school (K-12), with the aim of developing a means for educating the general public and especially school-age children about environmental health issues, including the risks and effects to human health from exposure to physical and chemical agents.

The NIEHS K-12 educational initiative is one of the first steps in helping citizens make informed choices regarding environmental health issues. Policy decision-making requires that citizens not only have a better scientific understanding of the environment and its influence on human health, but also a range of other information and skills, including, but not limited to: engaging in persuasive writing, understanding the roles and relationship between local, provincial, and national governments, and using probability and statistics to assess environmental risks. Therefore, environmental health offers an excellent opportunity for students to use and strengthen the knowledge and skills gained from typically disconnected subject areas.

The units and lessons described in the following chapters arise out of a NIEHS project, Environmental Health Science as an Integrative Context for Learning (EHSIC), that began in 2000 and funded efforts to implement cross-disciplinary environmental health curricula in science and non-science subject areas in order to improve "students' overall academic performance" (NIEHS 1999, 2). In particular, the initiative aimed for:

> The enhancement of critical thinking skills to develop a sense of personal involvement in the issues and challenges created by the linkage of human health with environmental factors; competence in identifying problems; assembling relevant data, arriving at solutions; deeper understanding of the concepts of environmental health sciences through applications of the scientific process; awareness of career opportunities in environmental health fields and the background essential to those careers; better performance on standardized tests through the use of EHSIC; and improved comprehension of all academic disciplines (NIEHS 1999, 5).

The NIEHS maintained that not only should environmental health be part of the curriculum, but also that the interdisciplinary nature of learning about environmental health would improve student learning across several subject areas, an argument with which we concur (Martina et al. 2009). Because learning about environmental health engages students in their local community and situates science within history, ethics, and community, students are more likely to be motivated and successful.

As described in Chap. 2: *The Science of Environmental Public Health*, environmental health, as defined by the World Health Organization, comprises those aspects of human health that are determined by physical, chemical, biological, social, and psychosocial factors in our environment. It involves the theory and practice of assessing, controlling, and preventing these factors in the environment that can potentially adversely affect the health of present as well as future generations.

However, as we describe below, we cannot always easily assess and distinguish environmental risks that may affect our health. For example, by eating fish we may expose ourselves to toxicants, in particular to those fish, at the top of the food chain, such as swordfish, shark and tuna, which are known to have high levels of mercury. Conversely, we are also informed that eating fish rich in omega-3 fatty acids are beneficial to our heart health. As the exact harm from the risk of long-term neurological dysfunction from the ingestion of mercury found in some fish and the benefits of omega-3 fatty acids in fish meals continues to be debated, we need to challenge students to examine the balance of this risk-benefit (Lynch et al. 2010).

Teaching about environmental health helps students understand that science is more than memorizing what has been determined to be true. Rather, what we know is less certain and limited. Science is active, not passive. For example, what we know about toxic levels and hazards to human health has changed over time. As we describe later, the natural element lead was once considered safe and was added to house paint and automobile fuel. Yet over time, our awareness of dangers from lead increased and even small amounts in the bloodstream are now seen as dangerous (Canfield et al. 2003; Lanphear et al. 2005) and it is well recognized there are no safe levels for lead. Teaching about environmental health, therefore, helps students understand that science, which is often presented as if we had a complete and true understanding of the world, is, instead often complex and contradictory (Brickhouse 1994; National Research Council 1996). In learning about environmental health, students are confronted with the fact that each day we learn about new risks—such as the danger of bisphenol-A or second hand cigarette smoke. The Centers for Disease Control's Fourth National Report on Human Exposure to Environmental Chemicals (2009) included exposure information to 75 additional chemicals, including, acrylamide, which is formed when food containing carbohydrates are cooked at high temperatures (e.g. fried foods).

Students will be confronted throughout their lives with new potential environmental health risks to which they will need to assess and respond. Because it is impossible to anticipate what new health risks will be identified, or even to respond to all that we currently know—we hope that we provide an approach that educators can adapt to evolving information.

On a broader level, we also encourage students to become part of what is a central question of our time: given that toxicants and toxins are both manufactured and occur naturally in the environment, how do we develop a world that best supports human and environmental needs, in which we meet our lifestyle needs, while limiting the introduction of toxicants into the environment?

Consequently, as we show in subsequent chapters, environmental health curriculum necessitates integrating other subjects, as students situate the scientific understanding of toxicants within a changing history of a community (the increase in airborne toxicants from increased vehicular traffic or from factories), or changing technologies (for example, the rise of pesticides and herbicides).

Therefore, many of the lessons and units in this book use approaches that encourage students to explore different viewpoints through debate and case studies, and most importantly, to critically evaluate the information they receive, and ask if these sources are valid and reliable, and evidenced-based and neutral or are they sponsored by or affiliated with a particular corporate or political group. Students are also asked to develop approaches to reducing their community's environmental health risks via both personal actions and public policies, such as presenting information to their parents, teachers, and students in their school or meet with public officials at the local and or state levels.

Furthermore, because environmental health is a subject that students have not studied in school, but are likely to have knowledge about through the media or discussions (for example, what are the health risks associated with local drinking water or air pollution?), we encourage teachers to first assess what students already know about a specific toxicant or toxicants in general. We suggest teachers begin by asking students: Are there things in the environment that can affect your health and what are they? How did you learn about this danger? Have you ever been affected by any environmental toxins or toxicants? Do you know anyone who has been affected by an environmental toxin or toxicant? Are there ways to prevent exposure? What are the short (acute) or long-term (chronic) health effects?

In Chap. 4 *Environmental Justice*, we describe a social studies teacher, Mike Fantauzzo who begins a unit by asking his secondary students to brainstorm examples of environmental health issues and public health. In the process, he asks them to evaluate whether the risks can be eliminated and the possible negative consequences of eliminating those risks. For example, if we eliminate vehicular traffic, how would we transport people or products? What would we need to change if we are to reduce the amount of air pollution via transportation?

Some of the units use Problem-Based Learning (PBL) as an approach. In PBL, students are first presented with a problem, then inquire, investigate, build solutions and take action. The projects that Mike Fantauzzo assigned to his twelfth-grade students are a form of PBL. His students are first presented with a hypothetical problem in the form of a letter from a fictional community member. In the example we provide, the letter is from "Ivanna Kleenup," a community resident who complains that her tap well-water has a foul smell to it, may be contaminated, and that this may be caused by either an adjacent dairy farm or a neighbor spraying pesticide on her roses. Next, the students are encouraged to use information on the internet that is valid and reliable, such as the U.S. Environmental Protection Agency and the European Environmental Agency (http://www.eea.europa.eu/data-and-maps/explore-interactive-maps). Through these and other resources they examine pesticide use and the subsequent risks to animals and humans, and offer an explanation for the water contamination. Lastly, the students proposed governmental policies aimed to decrease the health risks, and developed an educational component (typically an iMovie, pamphlet, or PowerPoint) to increase student awareness of the issue.

Similarly, Project EXCITE presented students with the following imaginary scenario:

> A large-scale diary farm was recently constructed and is now operating in rural northwest Ohio and another large-scale dairy farm is scheduled for construction in the near future. Others are likely to follow. Organized members of the local community have voiced great concern. What do you think should be done? As a citizen living in the community [student role], formulate an evidence-based position statement that indicates what you think should be done and develop a project to enact your position [student responsibility]. Be sure to take into account the needs and values associated with different stakeholders of the community [critical condition] (Haney 2004).

Problem-based learning, and other variants described here, supports the notion of science and society that we have described so far. As described by Checkly (1997):

> PBL is an instructional method that uses a real-world problem as the context for an in-depth investigation of core content.
>
> The problem that students tackle are ill-structured (initial information is left out and intentionally vague); they include just enough information to suggest how students should proceed through the investigation, but never enough information to enable students to solve the problem without further inquiry. These problems cannot be solved by using formulas—students must use the inquiry process and reasoning—and there may be more than one way to solve the problem.
>
> Teachers who use a problem-based learning approach become tutors or coaches, helping students understand their own thinking and guiding them as they search for new information.
>
> Through problem-based learning, students become better problem solvers because they hone skills such as reasoning, collaboration, and persistence in their self-directed search for solutions (p. 111).

The lessons, therefore, aim to teach students to think like scientists—the problems are complex, messy, and may have no obvious solution or several possible solutions—and introduce students to central biological and toxicological concepts. While much of the information we present in Chap. 2: *The Science of Environmental Public Health* may be new and challenging, particularly for some elementary and secondary students, and students have mastered them. In this section we will provide an overview of fundamental environmental health concepts and the ways in which they were presented by the author of this chapter to middle and high school teachers as summer institutes through the Maryland Public Television (MPT) Envirohealth Connections program. In fact, many of the lessons developed in this program are based on this material. Thus, our experience tells us that the concepts can be successfully learned by most teachers and taught to most students.

The curriculum ideas presented in Chap. 3, *Environmental Health Curriculum* focus on issues related to air, water, land, and food, which are the means or vectors through which we are exposed to hazardous agents. Potential hazardous agents include toxic chemicals and metals, pesticides, air pollutants, radiation, and microbes and their products. However, the chapter begins with a section on ways to teach children about *toxicology*, which is "the science that studies the harmful effects of over exposure to drugs, environmental contaminants, and naturally occurring substances found in food, water, air, and soil" (AMBIENT 2004, p. 1).

The module begins by introducing students to the concepts of exposure and dose. Hazards can enter the body through ingestion, inhalation and skin absorption. But, exposure alone does not inform us as to the risks imposed on the individual. Rather, one central issue is the dose received and the dose varies dependent on the concentration of the hazard, how often and how long the exposure occurred, and the properties of the toxicant.

Toxicologists often reiterate a sixteenth Century view of toxicology that "the dose makes the poison." However, recent (vom Saal and Peterson 2008) research reveals that it is more complicated. Because small doses may metabolize differently than large doses, smaller doses may have disproportionately negative effects.

Moreover, the timing of the dose can make a difference, in particular the developmental stage of humans and other organisms receiving the dose. For example, it matters when in the development of the fetus that exposure occurs (Axelrod et al. 2001), which has different effects over the developmental lifespan of that person (childhood to adolescence, to early adulthood, middle and old age).

Scientists are also beginning to examine the effects of toxicants in combination. Understanding the combination of exposures is much more realistic in exposure scenarios in that we and other living beings are typically not exposed to just one chemical compound but many and scientists are unsure how these combinations of chemical compounds are metabolized. Finally, there are also huge differences between acute exposures that are easier to diagnose in contrast to chronic exposures that are much more difficult to diagnose.

All of the lessons presented in the curriculum chapter essentially teach students about different toxicants in the environment. However, in the section on teaching toxicology, we begin with the most common and therefore accessible toxicants in a child's environment: those found in their home. These can include household and lawn and garden products, cleaning solutions, personal hygiene and cosmetics, and the improper disposal of electronic waste. Every day children are exposed to environmental toxicants and they need to be aware of the dangers and what they can do to reduce their risks. This includes eliminating the toxicant from their home and, if necessary finding alternative ways to meet the need.

After presenting some simple but effective ways that educators might teach about toxicology, toxicants, and their health effects, Chap. 3: *Environmental Health Curriculum* turns to providing overviews of several approaches to teaching about three vectors through which we are exposed to toxicants: soil, water, and air.

Common and controversial dangers from soil and water include pesticides and herbicides. We describe how teaching about pesticides and herbicides also provides an opportunity to teach about changing land use, and the water cycle. For example, students in *My Environment, My Health, My Choice* studied how pes-ticides can enter the water cycle from lawns and agricultural fields, and, therefore, become prevalent in streams and lakes, and eventually drinking water.

Food borne illnesses are also an area in which students should be interested and where they can affect either their own or their family's behaviors to reduce their chances of becoming ill. Therefore, the second area for which we provide some lesson ideas focuses on food preparation and safety.

Heavy metals pose a danger for children and we use lead as an exemplar. Lead is still found in paint, although to different degrees in different counties, in soil, again often depending on whether and when leaded gasoline/petrol was banned, and in electronic technological waste from cell phones, computers, and other appliances. Furthermore, our understanding of the toxicity of lead continues to change as our research reveals detrimental effects on humans at ever diminishing levels.

We also provide information on two other routes through which people are commonly exposed to heavy metals. One exposure sources is electronic waste, which poses a problem when it is not disposed of properly in landfills, or when it is sent to developing countries for recycling where workers, dismantling electronic waste for valuable metals, expose themselves to toxicants.

The last route in which we are exposed to toxicants is air. In this lesson we present a problem-based learning scenario in which children learn about the connection between air pollution and asthma. We also raise the issue of the detrimental health effects caused by cooking with wood and dung in developing countries and what schools might do to research and reduce those effects.

In all of the examples we suggest that such student driven investigations can facilitate the reduction of the use of toxic substances in our environment. In this way the lessons are not mere exercises conducted solely to complete an assignment, but can have real beneficial health effects on students' lives, the lives of their families, and their community. Toxicants in our environment pose real dangers that, if understood, can be alleviated or eliminated.

The last point suggests that we situate environmental health within the context of ethics and environmental justice. Students are often concerned with what is fair or just and, therefore, environmental justice is a topic that often resonates with them. The notion of environmental justice includes many concepts, such as how and who creates the environmental health dangers, and who is most likely to be vulnerable. The recent lead poisoning of more that 1,300 children in southeastern China's Hunan Province from a newly opened and unlicensed manganese smelter is an example of how one specific population, often rural, poor, and minority, are more likely to be affected than others (Wines 2009).

Moreover, environmental justice raises larger questions regarding the relationship between people and their environment. For example, is a healthy environment a right that everyone is entitled to, and, if so, what exactly does that mean? What is the relationship between people and the environment and how can we develop democratic and sustainable communities? What is the relationship between environmental sustainability, environmental health, and social justice?

Furthermore, issues regarding environmental justice are an excellent way to extend environmental health into the social sciences. Consequently, several of the sites developed units focusing on environmental justice and we describe two of those units in Chap. 4: *Environmental Justice*. First, we describe efforts by fifth grade students (ages 11–12) to research the dangers posed by pesticides and herbicides and their efforts to develop different media presentations to educate students, parents, teachers, and community members. We also detail students in a secondary government class who researched the dangers posed by pesticide use, lead poisoning, electronic waste, and hazardous materials transport, and then, after observing and learning more about local governance, developed public policies to reduce or eliminate the dangers. Lastly, we overview an ethics module created by AMBIENT in which students use case studies to examine toxicants in the air, water, soil, and food.

The AMBIENT module introduces several key concepts, such as the relationship between environmental degradation and human environmental health hazards, exposure data on routes of exposure, health effects through historical records and changes over time, and risk assessment. AMBIENT's ethical module provide background on ethics, ethical reasoning, ethical concepts and case studies of issues affecting air, soil, water, and living things. EnviroHealth introduces students to a range of issues, ranging from the effectiveness of hand washing, genetic testing, nuclear energy, antibiotic resistant bacteria, and the pros of banning smoking in public places.

Understanding the relationship between the environmental and human health, we argue, is one of the most important issues that our children will face. It is an area in which students can make a difference in reducing their own health risks and the risk of those in their family, school, and community. Unfortunately, questions regarding toxicants in our environment and associated health risks are rarely discussed in schools. Moreover, examining our relationship with our environment is central not only to our health, but also to the health of all living species. Environmental health is a crucial topic that must become an essential part of the curriculum. We hope that this volume contributes to that process.

References

AMBIENT (2004). *Toxicology— Relating to environment health.* Coral Gables FL: University of Miami.

Anto, J.M., Sunyer, J. (1990). Epidemiological studies of asthma epidemics in Barcelona. *Chest,* November Supplement:185s–189s.

Axelrod, D., Davis, D. L., Hajek, R., & Jones, L. (2001). It's time to rethink dose: the case for combining cancer and birth and developmental defects. *Environmental Health Perspectives, 109*(6), A246–A249.

Canfield, R. L., Henderson, C. R, Jr., Cory-Slechta, D. A., Cox, C., Jusko, T. A., & Lanphear, B. P. (2003). Intellectual impairment in children with blood level concentrations below 10ug per deciliter. *The New England Journal of Medicine, 348*(16), 1517–1526.

Checkly, K. (1997). Problem-based learning: the search for solutions to life's messy problems. *Curriculum Update, 1–3*(6–8), 109–113.

Eder, W., Ege, M. J., & Von Mutius, E. (2006). The asthma epidemic. *The New England Journal of Medicine, 355*(21), 2226–2235.

Haney, J. J. (2004). *The Odyssey Design and Implementation Process Project Excite: Taking Action, 2*(2), 2–3. Spring.

Kjellstrom, T., & Hick, N. V. (1998). Case X: Atmospheric fog in Greater London. In S. Markowitz (Ed.), *Problem Based Training Exercises for Environmental Epidemiology*. Geneva, Switzerland: Office of Global and Integrated Environmental Health, World Health Organization.

Lanphear, B. P., Hornung, R., Khoury, J., Yolton, K., Baghursh, P., Bellinger, D. C., et al. (2005). Low-level environmental lead exposure and children's intellectual function: An international pooled analysis. *Environmental Health Perspectives, 113*(7), 894–899.

Layton, L. (2009,July31). House approves overhaul of food safety regulations. *Washington Post*.

Lynch, M. L., Huang, L-A., Cox, C., Strain, J. J., Myers, G. J., Bonham, M. P., Shamlaye, C. F., Stokes-Riner, A., Wallance, J. M. W., Duffy, E. M., Clarkson, T. W., & Davidson, P. W. (2010). Varying coefficient function models to explore interactions between maternal nutritional status and prenatal methylmercury toxicity in the Seychelles Child Development Nutrition Study. Environmental Research http://dx.doi.org/10.1016/j.envres.2010.09.005

Martina, C. A., Hursh, D., & Markowitz, D. G. (2009). Contradictions in educational policy: implementing integrated problem-based curriculum in a high stakes environment. *Environmental Education Research, 15*(3), 279–297. PMCID: PMC108802.

National Institute of Environmental Health Sciences. (1999). *Environmental health science as an integrative context for learning*. Washington, D.C: National Institute of Environmental Health Sciences. RAF ES 99–011.

National Research Council. (1996). *National science education standards*. Washington, D.C.: National Academy Press.

vom Saal, F. S., & Peterson Myers, J. (2008). Bisphenol A and Risk of Metabolic Disorders. *Journal of the American Medical Association , 300*(11), 1353–1355.

Wines, M. (2009). Lead Sickens 1,300 Children in China, *The New York Times*, p. A-4.

The Science of Environmental Public Health

Michael A. Trush

Abstract In this chapter, we begin with how the World Health Organization (WHO) and the United States define environmental health. We then turn to describing environmental hazards and how to measure and assess environmental exposures. Next, we examine the process through which hazardous agents enter and influence body systems, and why individuals respond differently to toxic exposures. Then, we show the relationship between exposure and disease risk assessment. We conclude with case studies of two toxic agents—mercury and dioxin—that demonstrate how the concepts provided in this chapter can be applied.

Keywords Toxins · Toxicants · Dose response · Hazardous agents · Vectors · Air · Water · Soil

Introduction

In the Introduction, we briefly described some of the environmental health hazards that children and adults face: lead poisoning from paint and gasoline, pesticides and herbicides, air and water pollution, and food borne illnesses. Although these pose life-threatening dangers to humans, environmental health hazards are rarely taught in schools. Instead, students are more likely to learn about the effects that human's impose on the environment and its negative consequences for living things other than humans.

In the US, students are much more likely to be concerned about the effects of global warming on the plight of polar bears than that rising sea levels in Bangladesh are driving many who live close to sea level out of their homes. While

D. W. Hursh and C. A. Martina, *Teaching Environmental Health to Children*, SpringerBriefs in Environmental Science, DOI: 10.1007/978-94-007-1811-1_2,

we do not want to slight the plight of the polar bear, we contend that the 2 million people in developing countries, who each year die from inhaling the wood smoke from their cooking stoves (Legros et al. 2009), or the 700,000 people in China who die prematurely each year from excessive pollution (Barboza 2007) also deserve attention. We believe that it is essential that students learn about how the environment influences their own health and the health of those around them and what they might do to reduce or eliminate the risk and take steps to do so. Environmental health must become part of the elementary and secondary school curriculum.

However, understanding and responding to environmental health risks requires some understanding of environmental health science. In this chapter we begin with how environmental health is defined by the World Health Organization (WHO) and in US government documents. We then turn to describing environmental hazards and how to assess or measure environmental exposures. Next, we examine the process through which hazardous agents enter and influence body systems, and why individuals respond differently to toxic exposures. Then, we show the relationship between exposure and disease risk assessment. We conclude with case studies of two toxic agents—mercury and dioxin—that demonstrate how the concepts provided in this chapter can be applied.

The WHO defines environmental health as those aspects of human health that are determined by physical, chemical, biological, social and psychosocial factors in our environment (World Health Organization 2009). It involves the theory and practice of assessing, controlling and preventing those factors in the environment that can adversely affect the health of present and future generations. The significance of environmental health was further emphasized in the US document, *Healthy People 2010* (U.S. Department of Health and Human Services 2000) that aims to "Promote health for all through a healthy environment" (www.healthypeople.gov/Document). The major objectives with regard to the environment and human health focus on the following six categories: (1) outdoor air quality, (2) water quality, (3) toxics and waste, (4) healthy homes and communities, (5) infrastructure and surveillance, and (6) global environmental health. *Healthy People 2010* recognized that both our physical and social environments play critical roles in the health of both individuals and communities. It also recognized that an individual's biology (genetics and differences in susceptibility) is also an important determinant in how our physical and social environments influence our health.

From the above perspectives on environmental health, it is clear that environmental health encompasses all components of education from the biological and physical sciences, mathematics, and the social sciences. Incorporating environmental health concepts and concerns, therefore, becomes a way for schools to integrate across the K-12 curriculum. Students who understand the concepts of environmental health will also become better-informed citizens with regard to the importance of protecting environmental quality not only from the environmental perspective, but also how the environment impacts their health.

We Live in a World of Environmental Hazards

Environmental hazards include agents such as toxic chemicals, metals, pesticides, air pollutants, ionizing and non-ionizing radiation, and microbes and chemicals produced by them, such as the human carcinogen, aflatoxin, produced by the fungus *Aspergillus flavus*. These hazardous agents are in our air, food, water or soil that provide the vectors or carriers for exposure. Consequently, when we describe an environmental hazard, we are referring to either or both an agent or factor whose exposure may adversely affect health.

Exposure is the opportunity for contact between a hazardous agent and humans that could contribute to the development of environmentally associated morbidity (disease) and mortality. We are continually exposed to hundreds of environmental chemicals. In the National Geographic article, "The Pollution Within" (Duncan 2006), science writer, David Ewing Duncan, to draw attention to the intensity of our daily risks, voluntarily underwent expensive testing to identify up to 320 chemicals in his blood, including lead, mercury, dioxin, polychlorinated biphenyls (PCBs), and fire retardants. While none of the chemicals toxins were present at a level that created an immediate acute health danger to the author, the presence of these chemicals indicated that not only had he been exposed through various vectors or routes, but that these agents had been absorbed and retained in his body.

As the above article by Duncan demonstrates, we all encounter hazardous agents daily at home, school, work, during recreation, or while traveling. For example, while we are rightfully concerned with outdoor air pollution from cars, trucks, and factory smokestacks, ambient indoor air quality is where most everyday exposures to hazardous agents occurs (Ott and Roberts 1998). Quite simply, because we breathe (inhale), eat and drink (ingest), and touch surfaces (dermal), it is impossible not to be exposed to hazardous agents in our environment. Inhalation, ingestion, and dermal are the most common routes of exposure to hazardous environmental agents. Secondary maternal prenatal gestational exposures pose an additional route of significant concern for the development of the fetus and subsequent child. These exposures may have life-long health impact on the child throughout their development into adulthood. In addition, a growing concern is the low-dose exposure to hazardous agents to a developing fetus in utero (Colquhoun et al. 2009; Environmental Working Group 2005; The Endocrine Disruption Exchange: Critical Windows of Development, at http://www.criticalwindows.com/go display.php). For example, bisphenol-A (BPA) and phthalates, which are both commonly found in some baby bottles, processed foods and toys, pose a significant danger to young children. In the US, the food and drug administration (FDA) recently examined BPA health risks (Center for the Evaluation of Risks to Human Reproduction 2008; Grady 2010; Koch and Calafat 2009). The FDA stated that it "had some concern about the potential effects of BPA on the brain, behavior and prostate gland of fetuses, infant, and children." The potential human health effects surrounding chemicals in plastics, such as BPA and phthalates, continues to be a concern to government agencies such as NIEHS and EPA (Walsh 2010). It is also an example of the evolving awareness of dangerous toxins in our environment that none of the lessons described herein respond to the dangers of BPA. However, see the Center for the Evaluation of Risks to Human Reproduction (2008).

Exposure to hazardous agents can be either acute (short term) or chronic (long term). Both types of exposures can result in adverse health. While acute exposures to chemicals occur (for example, industrial accidents or carbon monoxide poisoning), resulting in readily recognizable adverse health effects, the main concern in environmental health are chronic exposures and low-dose exposures. Whether a harmful effect occurs or not following a chemical exposure depends upon factors that occur in the individual. These include how our bodies modify or biotransform (metabolize) these chemicals as well as how well our cells can repair damage from chemical insults. Both of these factors will be discussed in more detail below.

An important component of environment health is exposure assessment. Exposure assessment examines the manner in which humans come into contact with hazardous agents, the sources of these agents, and the concentration levels at these sources. It examines the magnitudes, frequencies and durations (acute or chronic) of such exposures. Determining the concentration of agents in the environment can be obtained in several ways. For example, release estimates from an industrial facility can be generated using industrial data and information on the production process. Industrial hygienists and environmental health engineers often do such assessments and use on-site measurements, mathematical equations, and computer models to estimate the path of a chemical in the environment over time.

From a community perspective, in the US industries are required to provide a toxic release inventory (TRI). Such information has proven to be useful to communities who are concerned about environmental justice. A useful exercise for students is to do a computer search using a zip code of interest to find TRI data with regard to local industries (http://www.epa.gov/tri/). The European Environmental Agency has numerous data resources, including maps at http://www.eea.europa.eu/data-and-maps/explore-interactive-maps.

The most accurate way to obtain environmental concentrations and human exposures is to perform an exposure monitoring study. This may entail putting carbon badges on individuals or backpacks with pumps to collect chemicals that can later be identified in the laboratory. Students have often participated in such exposure monitoring studies in their schools and communities (Northridge et al. 1999).

An important component of an exposure assessment study also involves estimating the level of contact of an individual within their environment. This includes collecting data on where individuals live (near a factory or highway) and work, as well as collecting data on their daily activities. Measurement of how much (dose) of an agent is absorbed (internalized) is becoming an increasingly important component of exposure assessment. Developing sensitive and accurate analytical chemical methods for the measurement of internal doses is a critical component of environmental health research.

Accurate exposure assessment is key to the risk assessment process and environmental epidemiology studies. The better the exposure assessment data, the more useful and informative are environmental epidemiology studies in applied environmental health.

"Does 'The Dose Make the Poison'?"

As revealed in Duncan's "The Pollution Within" above, humans are continually exposed to a spectrum of potentially hazardous chemicals. While many factors determine whether or not an adverse health effect occurs, a critical factor is the dose or amount.

Listed below, are some of the terms are used in environmental health that pertain to dose.

1. Exposure: Contact between an environmental hazard and a living organism.
2. Concentration: Amount of hazard per unit volume or mass in an environmental vector.
3. Internal dose/absorbed dose: The amount of an environmental hazard that crosses one or more of the body's membrane boundaries.
4. Biologically effective dose: Amount of the internal dose and/or its metabolite that reaches and interacts with a molecular target.

The relationship between dose and chemical toxicity was first developed by Paracelsus (1493–1541), who is considered the father of toxicology (Klaassen 2007). In a traditional broad sense, toxicology is defined as the science of "poisons." It grew out of the awareness during human history that toxins in plants and animal venoms could be used for suicides and executions, to achieve political gains and to do away with one's rivals. Paracelsus recognized that what determined whether something was a "poison" was dose and that anything could be a poison depending on the dose, hence, the commonly used phrase in toxicology "The dose makes the poison." This is particularly true of acute exposures and acute toxicity.

Drinking too much water or eating too much salt can be lethal. While this may be generally true, the research in this area is becoming increasing complex. See, for example, Myers and Hessler's "Does 'the dose make the poison?': Extensive results challenge a core assumption of toxicology" Environmental Health News (Myers and Hessler 2007).

To describe acute toxic effects, we use the term lethal dose 50 (LD_{50}) that represents the dosage administered (milligrams per kilogram body weight) that causes lethality in 50% of an exposed experimental animal group (five out of ten animals for example). The LD_{50} is a useful concept when comparing the acute toxicity of one chemical to another. While once common, lethality tests to determine an LD_{50} are becoming less common in toxicity testing. This has come about, in large part, due to public awareness about these types of tests and the number of animals required. Increasingly, toxicity testing is moving in the direction of in vitro tests, using cell and tissue culture, and computer modeling. Toxicity testing and hazard identification, along with dose–response relationships, are key components of the risk assessment process (Klaassen 2007).

Data from toxicity tests and experiments are plotted as a dose–response curve or graph to illustrate and describe the effect of exposure to an agent. Graphs can be plotted based on either individual responses (graded) or as population responses (quantal) that results in a sigmoidal cumulative percentage response curve (Klaassen 2007). On these dose–response graphs, dose is plotted on the *x*-axis and the response is on the *y*-axis. Many of the educational resources developed through NIEHS describe experiments students can do to generate data to plot dose–response curves. Having students plot such curves fits in well in both science and mathematics classes.

What Happens Following Exposure?

Exposure to a hazardous agent does not in of itself equate to toxicity or an adverse health effect. Hazardous agents first need to be absorbed or internalized. Absorption is one of a group of processes collectively referred to as *toxicokinetics*. Toxicokinetics represents the quantitation of the time course of a toxicant in the body following the processes of absorption, distribution, biotransformation (metabolism) and excretion or clearance from the body. It is a reflection of how the body handles toxicants as indicated by the plasma concentration of that toxicant determined at various time points.

The end result of these processes is a biologically effective dose of the toxicant. Toxicokinetics is expressed quantitatively by the determination of half-life, the time for the plasma concentration to decrease by one-half over a time course series of measurements. Hazardous agents can have a half-life ranging from minutes to years and is a useful assessment for comparing differences over time, both within an individual or between individuals. The time interval for taking medications (for example every 4 h) is based on the half-life of that medication determined in human volunteers participating in a clinical trial. However, the half-lives of both medications and environmental pollutants vary in humans with age. Half-lives tend to increase with aging.

The process of absorption requires the movement of a hazardous agent through cell membranes both at the site of exposure and at the various organs and tissues of the body. Once absorbed from its site of exposure, an agent is distributed throughout the body. The circulatory system is the primary way agents are distributed throughout the body, using blood plasma (liquid part) as the vehicle. Chemicals can move in and out of organs and back into the bloodstream and eventually be excreted. The primary routes of excretion of environmental agents are the feces and urine. These also represent the major quantitative routes of excretion. Chemicals can also be excreted in hair and fingernails, which has proven useful for assessing exposure to hazardous agents. Breast milk is another means by which chemicals can be transferred from a mother to a nursing infant. The dose of a chemical transferred via breast milk can be sufficient to elicit significant toxicity in the nursing infant.

Capillaries are the smallest blood vessels in an organ and are the site of absorption between the plasma and the cells of an organ. In terms of the distribution of chemicals, in certain organs there are anatomical barriers that reside at the capillaries that can restrict the distribution of chemicals. These include the blood–brain barrier, the blood–testis barrier, and the placenta. In terms of these anatomical barriers, the blood–brain barrier is the most restrictive. However, in humans, this barrier is not fully developed until around 10 years of age. Moreover, because the nerve connections in the brain are still developing, we need to be especially concerned regarding exposing children under the age of 10 to hazardous chemicals.

Lead and mercury are well known human developmental neurotoxicants. Lead poisoning remains especially widespread as children are exposed through lead found in the soil and homes as a result of the pervasive used of lead in gasoline, which while phased out in Europe and North America, is still used in some countries (Nriagu et al. 1996), and the addition of lead to house paint. Furthermore, children in some developing countries are exposed to lead as both children and adults engage in reclaiming used lead in lead-acid batteries, which is a lucrative and unregulated business (Potera 2009). Mercury also remains a problem as it is released into the atmosphere through coal-powered energy plants, altered through chemical or bacterial action into an organic form known as methylmercury, and then bioaccumulated in wildlife, particularly fish, that are consumed by humans (Environment Canada 2009). We discuss more fully in this introduction and the curriculum chapter.

Another factor that affects the half-life of an agent is whether the agent is stored in fat or bone. Compounds that are very fat (lipid) soluble can be stored for long periods of time and, therefore, tend to have very long half-lives. The compound 2,3,7,8-tetrachlorodibenzo-para dioxin (TCDD), also known as dioxin, is very lipid soluble and, consequently, has a half-life of around 10 years in humans.

Significantly, lead is chemically indistinguishable from calcium and is, therefore, stored in the bone. During human pregnancy, because of the increased demands by the body for calcium, lead is released from the bone. Furthermore, when older women experience osteoporosis, lead is released into the bloodstream. Therefore, in Chap. 3: 'Environmental Health Curriculum', we describe curriculum activities and student activism in which students learn about lead poisoning and what they might do to limit exposure.

As discussed above, urine and feces are the major quantitative routes of chemical excretion. In order for a chemical to be excreted efficiently via these routes, the chemicals must be water-soluble. This presents the body with a dilemma because in order to absorb most chemicals they must be lipid soluble, and to excrete them, they must be water-soluble. Accordingly, the body subjects these agents to enzymatic (enzymes are protein catalysts) reactions that convert the lipid soluble and non-polar forms to water-soluble and polar (charged or ionized) forms that can be excreted in the urine and feces.

This process is referred to as biotransformation. Two important enzymes involved in biotransformation are cytochrome P450 and glutathione-*S*-transferase. As a process, biotransformation should be viewed as a double-edged sword, having both a good and bad side. The good side is that through biotransformation lipid soluble compounds are converted to water-soluble metabolites which can be excreted and removed from the body. The bad side is that as a result of the interaction of chemicals with the enzymes involved in biotransformation, the chemical can be converted from a biologically un-reactive benign entity to a biologically reactive intermediate that can interact with biological molecules. Whether a chemical is converted to a biologically reactive intermediate is determined by its chemical structure and the molecular site on the chemical where biotransformation occurs. Essentially, biologically reactive intermediates are molecular entities that have an altered electron structure that is different from the parent (original) compound. In essence, chemicals prefer to be electronically neutral, thus biologically reactive intermediates interact with biological molecules in a manner that results in them becoming neutral. However, this interaction can result in a change in the structure and function of the biological molecule or target. By and large, the toxicity of most organic chemicals (those that contain carbon atoms) is due to their biotransformation to biologically reactive intermediates. This process is referred to as bioactivation. These intermediates fall into two broad categories: *electrophiles*, which are entities with a site of positive charge; and *free radicals*, which are entities that have a free or unpaired electron. Toxic inorganic metals, like lead and cadmium, are not biotransformed to biologically reactive intermediates. Thus, in contrast to organic molecules, the active form of metals that interact with molecular targets are the metals themselves.

How Do Chemicals Cause Toxicity?

How chemicals elicit toxicity can be stated as follows: The toxic action of a chemical results from the physical/chemical interaction of the active form with a *molecular target* within the living organism. For some chemicals, the active form is the agent you were originally exposed to, while for others it is a biologically reactive intermediate of that chemical.

Physical/chemical interaction pertains to the manner by which the active form reacts with a biological molecule. Some physical/chemical interactions are reversible, while others are irreversible. The interaction between an electrophile with its molecular target is irreversible because it forms a covalent bond where two molecules share electrons. This is the strongest type of chemical bond formed. Molecular targets refer to the biological molecules that react with the active form of the toxicant. Biological molecules fall into three categories: nucleic acids, such as DNA, proteins, and lipids (fat). Several examples of toxicant–active form interactions are illustrated below.

Some examples of toxicants and their molecular target
Carbon monoxide–Hemoglobin
Carbon tetrachloride–Membrane lipids
Dioxin–Arylhydrocarbon (Ah) receptor
Aflatoxin–DNA

Depending upon the category of the molecular target, different biological consequences can ensue. For example, modification of a cell's DNA may result in a mutation of a gene, in which the genetic code is altered. This genetically altered cell could serve as the initial cell in the development of a cancer.

The process in which an active form of a toxicant changes the molecular function, resulting in a change in organ function, such as the impact of a toxicant on DNA as described above, is toxicodynamics, or what the chemical does to us. First, all toxic responses start at the molecular level. Accordingly, the magnitude of the toxic response will be a function of the concentration of altered molecular targets, which depend on the concentration (biologically effective dose) of the active form of the toxicant at the cellular site where molecular targets are located. Therefore, the more molecular targets altered, the greater the biochemical changes in a target cell. The more cells affected in an organ/tissue, the greater the changes in the structure and function of that organ/tissue. Subsequently, if the structure/function of an organ/tissue is significantly altered, chemical-induced toxicity or disease will result. A difficult aspect in environmental health is determining the biologically effective dose and the extent of alterations in the molecular target population. With the exciting new developments in molecular biology and analytical instrumentation, environmental health scientists now have experimental approaches by which to address these issues.

Since we live in a world full of hazardous agents, and we are exposed to them on a daily basis, the above discussion may leave the impression that environmentally induced disease is inevitable. However, our cells are equipped with a number of defense molecules that counteract or neutralize active forms, as well as molecular mechanisms that can repair alterations in molecular targets.

Furthermore, we can improve our defense mechanisms by consuming foods, such as fruits and vegetables that contain a number of phytochemicals that are either in of themselves protective molecules or can cause the body to increase its natural defense and repair systems. For example, cruciferous vegetables, such as broccoli, contain high levels of the phytochemical sulphorophane. Among fruits, blueberries contain abundant levels of protective antioxidant phytochemicals. We are still learning which phytochemicals many fruits and vegetables contain and their biological activity and potency. For this reason it is critically important to stress to students the importance of eating a diet rich in a variety of fruits and vegetables.

What Accounts for Differences in Response Between Individuals?

Understanding that diet can improve your body's ability to resist toxic agents is one component in appreciating that individuals differ in their susceptibility. Given the same exposure, different individuals may exhibit different responses from that exposure.

Factors that contribute to differences in susceptibility include age, sex, diet, lifestyle and the presence of other diseases. The most basic and important factor in individual susceptibility is our genetics. While genetics contributes to our susceptibility, susceptibility only begins with exposure to an environmental hazard, which is referred to as a gene-environment interaction. As discussed in Chap. 4: 'Working for Environmental Justice Through Environmental Health', some groups of people, typically those who are poorer, tend to be more likely to be exposed to environmental hazards because of where they live. This is often the basis for environmental injustice.

Except for identical twins, we all have different genetic make-ups. However, even though identical twins have the same genetic makeup, the expression of their genes may differ (Fraga et al. 2005). Over the last several years, we have come to appreciate that gene expression is regulated, not through a mutation, but through a process referred to as epigenetics [see Brown's "Environment Becomes Heredity" (2008)]. Epigenetics involves the addition or removal of methyl (CH_3) groups to specific gene regions, CpG islands, or the acetylation and metylation of DNA bound histones, the proteins surrounding DNA. As such, genes can be expressed or be shut off. Epigenetics is a part of normal biology. The January 6, 2010 issue of *Time* magazine had several interesting and useful articles describing the role of epigenetics in human biology. It is becoming increasingly apparent that components in our diet and exposure to environmental chemicals can regulate genes through epigenetics. Such epigenetie regulation can occur in utero, which can result in a process referred to as imprinting (Dolinoy and Jirtle 2008).

In terms of toxickinetic processes, our genes determine and regulate the levels and amounts of enzymes involved in biotransformation. Individuals can inherit different forms of the same biotransformation enzyme. Furthermore, these different forms will have different catalytic activities (rate of catalysis), which is termed a polymorphism.

Given that so many chemicals depend upon bioactivation to elicit toxicity, polymorphisms in these enzymes represent a major susceptibility factor to environmental hazards.

Again, our individual choices can affect the frequency and kind of enzymes involved in activation. For example, exposure to chemicals in our diet, charcoal-broiled meat for example, and environmental chemicals, such as those in cigarette smoke, can increase the amount of enzymes involved in bioactivation, resulting in enhanced bioactivation of chemicals to biologically reactive intermediates, and leading to more molecular target alterations or disease. Conversely, phytochemicals in fruits and vegetables can increase biotransformation enzymes, consequently increasing the detoxification of biologically reactive intermediates, and resulting in less molecular target alterations or less risk for chemical-induced disease. The molecular and biochemical process by which increases in the levels of biotransformation enzymes occurs is termed induction.

The ability to induce our biotransformation enzymes and/or cellular protective systems has been found to be dependent on the interaction of chemicals with receptor/sensor target molecules. Two such receptor/sensor target molecules found in the cytoplasm of cells are the arylhydrocarbon receptor (Ah receptor) and nuclear factor E-2 related factor 2 (Nrf2). The Ah receptor, as well as its associated companion molecules, is genetically determined. Components in cigarette smoke, polycylic aromatic hydrocarbon (PAHs) for example, can interact with the Ah receptor. Through interacting with the Ah receptor, PAHs can induce enzymes involved their own bioactivation to electrophiles. Both animal- and human-based studies have shown that the risk for PAH-induced toxicity is genetically linked. Many of the beneficial phytochemicals found in fruits and vegetables interact with Nrf2. Induction through the Nrf2 mechanism has been shown to be protective against the toxicity of many chemicals, including PAHs (Osburn and Kensler 2008).

Individuals with xerodema pigmentosa (XP) are another example of a genetic–environment interaction. Xerodema pigmentosa is a genetic disorder leading to the deficient repair of DNA damage caused by ultraviolet light exposure. As a result of sunlight exposure, individuals with XP are at a much greater risk to develop several types of skin cancer. In very severe cases of XP, it is not only necessary for these individuals to avoid exposure to sunlight, but moonlight, as well.

Of course, while those with XP have to avoid exposure to sunlight, everyone, especially children, should limit their exposure to the sun to minimize the risk of skin cancers. Risk factors for susceptibility to skin cancer include: fair skin; a history of sunburn; excessive sun exposure; living in a sunny or high-altitude climate; having moles; and a family history of skin cancer. Because of their geographic location near the equator, Australians have a very high exposure to sunlight and thus a high incidence of skin cancer. Accordingly, to reduce exposure to sunlight, Australia has instituted a policy known as Slip, Slap, Slop: slip on a shirt, slap on a hat and slop on sunscreen. Not only is it critical for children and teenagers to be aware of the implications of exposure to sunlight for the development of skin cancer, but also for the development of cataracts later in life. Because of the concern of exposure to ultraviolet radiation, there is also concern associated with exposure in tanning booths, particularly among teenagers.

Another factor that results in individual differences is the presence of underlying disease, which can also contribute to increased susceptibility to response to air pollutants. For example, individuals with angina are particularly susceptible to the effects of inhaling carbon monoxide, a criteria air pollutant (i.e., in the US, a pollution for which there are national air quality standards). Carbon monoxide (the active form) targets hemoglobin, thereby altering the blood's ability to carry and release oxygen to organs. Individuals with angina are more susceptible to heart pain following exposure to carbon monoxide because, compared to the average individual, they deliver less blood and oxygen to the heart. Consequently, the government used the response of these patients to determine the regulatory criteria for carbon monoxide exposure.

In contrast, the increase in asthma rates increasing worldwide (Woolcock and Peat 1997) is thought to have more to do with environmental changes. Asthma is characterized by lung inflammation and increasing airway hyperactivity resulting in constricted airways and decreased ability to deliver oxygen in the lungs (Hargreave and Nair 2009). Many environmental agents can serve as asthma triggers, including criteria air pollutants, such as ozone and particulate matter, and cigarette smoke. Many of these asthma triggers are located in indoor air. Thus, those with asthma are much more sensitive to lower levels of pollutants than non-asthmatic individuals. However, individuals and families can take actions to reduce the their changes of inciting asthma, including staying indoors when ozone levels are high, avoiding particulate matter and cigarette smoke.

As described previously, because the blood–brain barrier is not yet fully developed, children are more susceptible to a number of hazardous agents. If a mother has lead stored in their bone, exposure to lead can begin in utero. As will be described more fully later, the dangers of in utero exposure were highlighted following the effects observed to organic mercury in Minamata, Japan. Another condition observed in some children born prematurely is the development of jaundice due to the buildup of bilirubin. Bilirubin is a product of the body's breakdown of heme-(iron) continuing molecules. Bilirubin relies upon biotransformation to a water-soluble metabolite in order to he excreted. If the enzyme that catalyzes this process is not present when a child is born, then bilirubin cannot be excreted and will accumulate in the body. Accordingly, light therapy is used to breakdown bilirubin until the development of this enzyme occurs.

On the other end of the age spectrum are the elderly. As mentioned earlier, in the elderly, the half-lives of a number of medications are known to be increased as compared to in younger individuals. This is believed to be due to factors that occur with human aging: a decrease in biotransformation ability and a decrease in kidney function. The latter may be due to accumulation of metals during a person's life.

One of the goals of environmental health research is to identify the susceptible individuals within the population and to understand the molecular and biochemical basis for these susceptibilities. In this regard, the regulation of pollutants is set at the levels of response of susceptible sub-populations as indicated above for carbon monoxide, individuals with underlying angina, and ozone, children and individuals with underlying lung disease.

We can summarize that because of genetics, body size, and age that individuals vary in their susceptibility to toxic agents and that increasing our understanding of these differences is important to improving our health. Furthermore, because of the individual environment interaction, it is important that we understand the ways in which different environmental agents can serve as triggers in particularly susceptible populations.

Linking Environmental Exposures to Human Disease: Environmental Epidemiology

Much of the discussion up until now has focused on toxicology, one of the two fields that environmental health draws upon. The other major public health field it draws upon is epidemiology. As a discipline, epidemiology deals with the distribution of diseases and relies heavily upon the use of mathematics and biostatistics. Epidemiology focuses on population studies rather than on individuals, and like most disciplines, it is divided into sub-fields. Environmental epidemiology refers to that component of epidemiology that examines the associations between environmental factors, and exposures and the diseases and health outcomes linked to these exposures (Friis 2007). Environmental epidemiology studies can aid other environmental health scientists in the formulation of hypotheses. In this section we will describe the various kinds of epidemiological studies, how they are performed, and difficulties that are faced.

Two measures of disease frequently used in epidemiology are prevalence and incidence. Prevalence refers to the number of existing cases of a disease, health condition or deaths in a population at some designated time relative to the total number in the population from which the cases in the study are derived. Incidence refers to the occurrence of new cases or mortality in the total population at risk within a defined time period observation.

The two common experimental designs used in epidemiology studies are case-control study and cohort study. In case-control studies, the subjects who participate are defined on the basis of the presence or absence of a disease outcome of interest, such as whether or not they have lead poisoning or asthma. As much as possible, the cases and controls are matched by age, sex, race or smoking history. Sources available for the selection of cases and controls may be individuals who work in the same industry, clinics, or disease registries. The outcome of association for a case-control study is known as the odds ratio (OR). An odds ratio greater than one (>1) suggests an association.

In an environmental epidemiology case-control study, the researcher must first assess to what each subject have been exposed. The researcher may interview subjects regarding what they recall of their past exposure or diet history. Consequently, the accuracy of their recall becomes important. In some cases, records may have been kept that document exposure during a particular time of interest, such as working in a particular industry. In such studies, the odds ratio refers to a ratio of odds in favor of exposure among those in the disease group, as compared to the odds in favor of exposure among the controls. An odds ratio greater than one (>1) indicates a positive association between the exposure of interest and disease. A difficulty in environmental epidemiology arises when the odds ratio is only slightly greater than one (>1), therefore requiring careful statistical analysis.

The second major epidemiological design is the cohort study in which subjects are classified according to their exposure of interest and then observed over a period of time to see the occurrence of new disease cases (incidence). Those in whom the disease occurs are then compared to those in whom it does not, with the latter group, therefore, forming the control group. Since in this design individuals are followed over time, cohort studies are longitudinal. The design of a cohort study may be prospective (looking forward) or retrospective (looking backward). The measure of association determined by a cohort study is called relative risk (RR). Relative risk is the ratio of the incidence of a disease outcome in an exposed group to the incidence rate of the disease in a non-exposed group. Mathematically, RR provides a ratio of two risks. Again, a relative risk greater than one (>1) indicates that the risk of disease is greater in the exposed group than in the non-exposed group. Sometimes a relative risk of less than one (<1) occurs. This would indicate that the exposure was associated with a protective effect. Like the cohort study design, statistical analysis and significance is essential in determining how strong the association is between the exposure and disease.

In designing environmental epidemiology studies, there are many factors that need to be incorporated and accounted for in the study design to ensure that any associations made are scientifically valid. In environmental epidemiology studies, there are also some limitations that need to be appreciated. These include the long latency periods between the exposure and disease appearance, such as cancer. Cancer is a multistage process and the development of this disease requires decades to develop. As described previously, exposure to hazardous agents may begin in utero, but the appearance of the agent-associated disease may not appear until adulthood. The recognition of this possibility is a current major focus in environmental health research. As discussed above, this possibility and concern has arisen from studies involving in utero exposures in experimental animals and occurs primarily through epigenetic mechanisms.

Another problem that environmental epidemiologists face is the low incidence and prevalence of certain diseases, which makes it difficult to obtain a sufficient number of cases to achieve statistical significance in an epidemiological study. This sometimes has practical and frustrating implications for communities that members express concern about a disease and chemical exposure in their neighborhood. While the number of cases may appear high to the community, there may not be a sufficient number of cases to perform an epidemiological study.

For environmental epidemiology to be accurate, a good exposure assessment is a critical component. When high levels of exposures to hazardous agents have been documented, establishing a relationship between exposure and disease becomes a much easier task. However, many environmental exposures occur at low levels, and there may have been a combination of differing multiple toxic exposures. Thus, establishing the relationship between the exposure to a particular chemical and disease becomes much more difficult.

In contrast to the two approaches described so far, increasingly epidemiological studies incorporate lab based molecular and analytical techniques. While epidemiology is not primarily a laboratory-based study, more and more epidemiology studies are incorporating molecular and analytical techniques or molecular epidemiology. Molecular techniques have been incorporated in studies in rural China linking exposure to aflatoxin, a mold-derived carcinogen, to the incidence and prevalence of liver cancer (Qian et al. 1994). Likewise, molecular techniques for assessing DNA repair have been useful in linking the risk of skin cancer with environmental exposures (Matta et al. 2003; Wei et al. 1994).

For example, in Taiwan, individuals are exposed to arsenic, a carcinogenic metal, from standing in water while working in rice fields (Grossman, unpublished observations). Arsenic-related cancer is also of current concern in Bangladesh due to the tub-wells used inresponse to the increased need for drinking water.

Connecting Exposure to Disease Risk: Environmental Risk Assessment and Communication

Life is full of risks, and in short, involves the likelihood of experiencing an adverse effect or event. For example, any time we drive a vehicle, we are at risk for getting involved in an accident. The assessment of risk is a qualitative or quantitative estimation of the likelihood of the occurrence of an adverse effect.

In terms of risk from synthetic agents and natural elements, on the one end of the spectrum we have exposure to hazardous agents, and on the other end we have the potential for the development of toxicity/disease. How do we connect them? Scientifically, this is done primarily using the disciplines of toxicology, exposure assessment, and epidemiology. Mechanistically, it is accomplished through risk assessment, which is the process of determining risks to health, attributable to hazardous agents in the environment. In environmental health risk assessment, there is a prescribed four-step process that includes: exposure assessment, hazard identification, dose–response assessment, and risk characterization. If a risk is identified, then some government agency, such as the EPA, may undertake steps to either eliminate the risk or lower their exposure to some level of risk deemed acceptable for the population. This is known as risk management and occurs through regulatory agencies passing regulations or laws, such as the Clean Air Act. Subsequently, this information needs to be presented and discussed with appropriate stakeholders, including communities, through risk communication.

The more knowledgeable citizens and communities are aware of these laws and regulations, the better they can ensure that these regulations are being enforced by the appropriate agency, whether state or federal. While risk management is a formal governmental process aimed at protecting the overall population, this should not stop individuals from taking steps to reduce or manage risks they have control over, such as reducing their exposure to ultraviolet radiation, controlling the environment within their homes, and maintaining a healthy diet.

Hazard identification comes from data generated from in vitro and in vivo animal toxicity studies, and epidemiology studies that examine the evidence that associates exposure to an agent with its toxicity. Within government and regulatory agencies, there are prescribed experimental protocols for assessing toxicity. Such assessments can range from acute to chronic tests. Moreover, there are prescribed protocols for performing reproductive toxicity tests and identifying if a chemical is a carcinogen (cancer-causing agent).

The development and utilization of toxicity tests is governed by the three R's: reduction, refinement, and replacement. Reduction in the number of animals used in toxicity testing; refinement in the way the animal tests are conducted to get the same answer, and replacement, if possible, of the animal test with an in vitro method.

Originally, the experimental endpoint for acute toxicity testing began with determining an LD_{50}, or lethal dose. However, fewer LD_{50} tests are being performed as a result of a greater awareness and incorporation of the three R's. A central limitation of animal toxicity tests is that these results must be extrapolated from animals to humans. Similarly, animal tests are often conducted at doses much higher than that to which humans are typically exposed. This requires high dose to low extrapolation in the risk characterization.

Rather than conducting research on humans or animals, scientists are particularly interested in assessing chemical toxicity through in vitro tests, or tests in a lab (Goldberg and Hartung 2006). Developing appropriate mechanistic based in vitro tests are especially important given that there are approximately 80,000–100,000 chemicals registered for use in products, many of which have not undergone extensive toxicity testing. We do not have the financial resources to test all of these chemicals with in vivo animal studies. Most of these chemicals were not intended for administration to humans. However, we may be exposed to them during their manufacture, distribution, use and disposal.

In contrast, pharmaceuticals are designed for human administration. Following both in vitro and in vivo animal testing, in the US the safety of drugs is evaluated through randomized clinical trials regulated by the FDA. The earlier human trials assess the toxicokinetics in humans, as well as the safety and efficacy (does it function pharmacologically as intended?) of the drug.

Another distinction between environmental chemicals and drugs is that the active form of environmental agents is generally a biologically reactive intermediate resulting from bioactivation, while for pharmaceutical agents the active form is the drug given. For many drugs, their toxicity is an extension of its intended pharmacologic activity. However, there are a number of drugs that undergo bioactivation to a reactive intermediate. Acetominophen, the active compound in Tylenol, is a classic example in toxicology of a drug whose liver toxicity results from its biotransformation to an electrophilic reactive intermediate that attacks cellular proteins leading to the death of liver cells.

Risk characterization involves doing extensive dose–response evaluation. Dose–response studies are designed to ascertain certain values: a threshold, a lowest observed adverse effect level (LOAEL), and a no-observed adverse effect level (NOAEL). The threshold is the lowest dose of a chemical at which a particular response would first occur. Risk characterization involves doing extensive dose–response evaluation. Dose–response studies are designed to ascertain certain values: a threshold, a LOAEL, and a NOAEL ([CM1] <#_msocom_1>). On a dose–response curve, the threshold dose is where the experimentally-derived dose–response curve intersects (meets) the x-axis. For both experimental and safety reasons chemical carcinogens are considered to not have a dose threshold, while for all other chemical toxicities, a threshold dose can be determined graphically.

The exposure assessment component of risk characterization involves estimating exposures in defined populations; (a) what exposures are anticipated or experienced, and (b) the exposure scenarios. If human epidemiological data exists, such information is considered in the risk characterization and supersedes data from animal toxicity studies. With human epidemiological data, the need for animal to human data extrapolation and high dose to low extrapolation is lessened. All of this data is brought together in the risk characterization to estimate occurrence of the adverse effect in a given population.

Based on the risk characterization, the development, evaluation and implementation of regulatory options aimed at reducing risk and control are put forth. This falls under risk management. In the US, the major agency involved in developing such regulations is the EPA. Occupational safety and health falls under the National Institute for Occupational Safety and Health (NIOSH). Many states have their own agencies for setting environmental health regulations. California, for example, has its own Environmental Protection Agency and an Air Resources Board.

The European Union (EU) has established a comprehensive system of environmental protection. Within the EU is the European Environment Agency (EEA) in Copenhagen. In 2007, the EU passed a new European Community Regulation, REACH, which stands for Registration, Evaluation, Authorization, and Restriction of Chemical substances, that places greater responsibility on manufacturers and industries to manage the risk from chemicals that they either manufacture or use in their products, including requiring that any company in the EU that sells chemicals and products containing chemicals provide adequate safety information. The WHO also provides leadership in decreasing adverse health outcomes associated with environmental pollutants and industrial development. Environmental health education is essential if citizens are to become more aware of global environmental health issues and to support and advocate for environmental health policies and reforms.

Risk communication involves providing risk assessment information and education to various stakeholders, which includes legislators, students, parents, and other residents. An informed citizenry is critical for protecting human health and addressing environmental justice from the local to the global levels. Risk assessment and environmental health policy development integrates all facets of the education spectrum: risk characterization in the biological and physical sciences and mathematics classes; risk management in classes that deal with how government agencies develop laws and policies; and risk communication in classes that teach communication skills. Effective risk communication relies upon effective writing and speech, and various visual techniques such as art and graphic design.

Putting it all Together Through Case Studies: Mercury and Dioxin as Examples

So far, we have tried to provide the reader with a basic understanding of environmental health sciences. A good way to bring these concepts together for teaching purposes is through a case study dealing with a specific issue. For this purpose, in addition to the various materials described in this book, a very good website is *Toxtown* (http://toxtown.nlm.nih.gov). *Toxtown* has useful materials on environmental concerns and toxic chemicals where people live, work, and play. The website is interactive, with a teacher section containing classroom activities, interactive resources, and a description of careers in environmental health. The same materials are also presented in Spanish.

Case Study on Mercury

Mercury is a naturally occurring element found in the earth and is designated by the chemical symbol Hg. At room temperature, it is the only element that exists as a liquid. It is in this form that individuals are most familiar with mercury, as it was widely used in thermometers, and barometers, and in instruments for measuring blood pressure. However, elemental mercury is not the only chemical form of this element. There are also mercury salts, which are used in antiseptics, and organic (carbon-containing) forms of mercury, methyl mercury (CH_3Hg) and phenylmercury. Phenylmercury was used in fungicides and the inappropriate mixing of this fungicide with agricultural seeds resulted in contaminating food that led to over 500 deaths in Iraq in 1971 and 1972. When we talk about the toxicology of mercury, we need to distinguish what form of mercury we are discussing, as its particular chemical form determines its toxicokinetics.

Elemental mercury is not toxic when ingested. It is not very lipid soluble and, thus, is not readily absorbed from the gastrointestinal tract. However, elemental mercury easily becomes a gas upon heating and because elemental mercury has a high affinity for gold, miners would use it to separate gold from rock by heating the rocks, causing the mercury to evaporate, and leaving the gold behind. However, inhaling mercury vapors can be very toxic to the lungs.

Unfortunately, mercury is still being used in gold mining around the world. Larmer, in "The Price of Gold" (2009), describes not only the devastating environmental effects of gold mining, but also the continued use of mercury to extract gold from rock by many of the small-scale artisanal mines. Around the world these miners are being exposed to mercury vapors, while waste mercury is entering river sediments and eventually poisoning the local food chain (Matchaba-Hove et al. 2001).

Elemental mercury enters the atmosphere through at least two common means. Because coal contains elemental mercury, burning coal to generate electricity is the major source of elemental mercury release to the environment. Also, incinerating mercury-containing batteries, thermometers, and other medical devices releases mercury into the atmosphere, which is why these products should not be disposed of in a manner where they will either be incinerated or put into landfills. When elemental mercury settles out of the atmosphere and enters aquatic environments, bacteria can biotransform elemental mercury into methyl mercury, an organic form. Interestingly, humans, unlike bacteria, lack the enzyme that can biotransform mercury into methlymercury.

In aquatic environments, these bacteria are part of the plankton, which is a major food source for small fish. Small fish are the food source for larger fish, all the way up to the food chain to the largest fish. Since methyl mercury is very lipid soluble, it can bioaccumulate and its concentration is magnified up the food chain. Thus, large fish, many of which are food sources for humans, such as tuna and swordfish, contain the highest concentrations of methyl mercury. When ingested by humans, methyl mercury is rapidly absorbed in the gastrointestinal tract, and being lipid soluble, easily passes the blood–brain barrier.

The most famous incident of methyl mercury poisoning occurred in Minimata Bay, Japan in 1956, in a community where most families fished for a living (Allchin 1999). After World War II, the Chisso factory began to manufacture acetaldehyde, which required elemental mercury for its synthesis. Mercury-containing effluents from the factory were deposited directly in Minimata Bay, and the fish in the bay became contaminated with methyl mercury. Eventually, methyl mercury poisoning began to appear. While adults experienced some health effects, the major effects were observed in newborn children, whose mothers ate contaminated fish while pregnant. Methyl mercury easily passed the placenta, and the developing fetus became a reservoir for the methyl mercury. The methyl mercury targeted the developing central nervous system leading to a number of neurological problems.

Initially, officials and medical directors of the Chisso factory denied that they were the source of mercury contamination in Minimata Bay. Similarly, the government failed to cite the Chisso factory as the source of the problem. However, researchers established the association between the health effects and methyl mercury exposure as a result of an unique part of the Japanese culture: after children were born, parents kept the child's umbilical cord. Analysis of the blood in the umbilical cords revealed a direct correlation between the increase in methyl mercury with the manufacture and release of acetaldehyde by the Chisso factory. There was little, if any, methyl mercury in cord blood prior to the Chisso factory manufacturing and releasing its effluent into Minimata Bay. Although the contamination occurred over five decades ago, waters and fish in Minimata Bay are still contaminated with mercury.

Moreover, eating fish contaminated with methyl mercury is still a world-wide concern. In the US, the FDA recommends that pregnant women, in particular, restrict their intake of fish because of the susceptibility of the developing fetus. The FDA also recommends that the general public reduce their ingestion of fish known to be high in methyl mercury content. Fish that have levels of mercury include swordfish, tuna, shark, and tilefish. Those that generally have low levels of mercury include salmon, trout, cod, flounder, and catfish. Because fish contain omega-3-fatty acids, this presents an interesting risk–benefit dichotomy since eating a diet containing fish is recommended for good cardiovascular and neurological health. Whether to eat fish and how often brings us back to the ideas discussed earlier about dose–response and dose-timing relationships.

Mercury exposure and human health has been raised in several other circumstances. Because of its affinity for metals, mercury is a component of silver tooth fillings known as amalgams. However, mercury is released with chewing. Scientific evidence indicates that this exposure does not represent a health problem (Porter 2010). Fortunately, because of the fluoridation of water, tooth decay is less common, and the use of amalgams has decreased significantly.

Mercury is also a component of thermosil, a preservative that was used in vaccines. Because of mercury's ability to cause development neurotoxicity, a significant concern was raised about the relationship between vaccinations and the dramatic increase in autism. Many studies have examined this relationship and despite a recent government report concluding that there is no relationship, significant doubt and concern still exists (Dyer 2010).

Case Study on Dioxin

The compound 2,3,7,8-TCDD is the chemical commonly referred to as dioxin. Dioxin belongs to a family of dioxin-like compounds that are un-intentional by-products of certain industrial and non-industrial processes that utilize halogen molecules like chlorine. Dioxins are also produced when wood pulp is bleached during the manufacturing of white paper. TCDD became well known during the Vietnam War, as it was a contaminant of an herbicide (weed-killer), Agent Orange, used to defoliate trees so as to more easily observe troop movements. Agent Orange was composed of two herbicides 2,4-0 and 2,4,5-T, and during their manufacture, TCDD was produced. As a chemical, TCDD has never been produced and sold as a commercial product. A large number of epidemiologic studies have concluded that there is an association between Vietnam veterans exposed to Agent Orange and the development of several forms of cancer. As a country, Vietnam is still heavily contaminated with TCDD (Nhu et al. 2009).

There have been a number of industrial accidents where workers and nearby residents were exposed to dioxin. The last major accident that resulted in human exposure to TCDD occurred in Sveso, Italy on July 10, 1976. According to researchers following the residents of Sveso, about 37,000 people are believed to have been exposed to the chemicals. The one health effect common to all of the reported industrial accidents resulting in TCDD exposure is the development of chloracne, a skin condition characterized by a very severe and disfiguring type of acne. This is the same effect seen in Viktor Yuchshenko, who, when he was campaigning for president of the Ukraine in 2004, was deliberately poisoned with TCDD.

Because of its high lipid solubility, dioxin bioaccumulates up the food chain. Thus, for the general population, human exposure to dioxins is through our diet. Dioxin enters the atmosphere when industrial and municipal waste, particularly plastics containing chlorine, are not incinerated at the proper temperature. Following release into the atmosphere, TCDD settles out and becomes part of the food chain. TCDD is transmitted to cows through the eating of TCDD-contaminated grass and then accumulates primarily in the fat or milk of cows.

After drinking cow milk, TCDD is readily absorbed from the gastrointestinal track and distributed throughout the body. The structure of TCDD is not amenable to biotransformation, so this highly lipid soluble compound remains lipid soluble. Consequently, TCDD is slowly excreted in the urine, thereby, accounting for its half-life of approximately 10 years in human beings.

Because it is so lipid soluble, TCDD easily penetrates into cells. Once in the cytoplasm (water portion of cells) of a cell, TCDD interacts with its molecular target, the Ah receptor. As mentioned earlier, the Ah receptor acts as a transcription factor leading to the induction of gene transcription and expression. Approximately 200 genes have been shown to be regulated through this mechanism.

The acute exposure of experimental animals to TCDD can be lethal. Lethal doses of TCDD in various species ranges from the microgram per kilogram to the low milligram range per kilogram range. Because such low doses can be lethal, TCDD is considered one of the most toxic chemicals of concern to human health. Interestingly, no human fatalities have been reported from acute TCDD exposure.

TCDD has a number of other effects in experimental animals. These include: chloracne, liver damage, endocrine effects, immune alterations, and reproductive and development effects. As mentioned earlier, chloracne has been clearly demonstrated following human exposure. Some epidemiological studies have exhibited a dose–response relationship. While the studies of Vietnam veterans have shown an association with Agent Orange exposure and the development of some forms of cancer, epidemiological studies with other populations have suggested an association with the development of cancer of the liver, but the OR was just slightly greater than one. Other epidemiological studies have hinted at an association between TCDD and other human diseases, but the OR of these studies are often around one (1) and either barely or not statistically significant.

Because TCDD is so prevalent and persistent in the environment, it is likely that most human beings have some amount of body burden of this compound, particularly arctic populations that eat sea mammals. Dioxins are very stable in the environment, and when released into the atmosphere can travel over vast areas before settling on the earth's surface or in aquatic environments becoming part of the sediments. Realizing that these compounds bioaccumulate up the food chain, we can reduce all exposures by watching the amount of fat in our diet. A good suggestion is to cut the fat from meats and to drink low-fat milk.

Dioxins belong to a group of compounds referred to as persistent organic pollutants or POPs. These include other dioxin-like compounds, such as the polychlorinated dibenzofurans (PCDFs) and certain PCBs. Like dioxins, these other POPs are very persistent in the environment and can travel long distances. Many of these chemicals also target the Ah receptor. Of the 419 dioxin-like compounds, TCDD interacts with the Ah receptor with the highest binding affinity, making it the most potent toxicologically. Another chemically-related POP is the polybrominated diethylethers (PBDEs), which are flame retardants used in furniture fabrics and electronics. Many European countries have restricted the use of PBDEs because of their appearance in breast milk and the concern for child health.

Furthermore, human exposure to dioxins and PBDEs are occurring in different parts of the globe, such as Africa and China, due to the burning and salvaging of electronic waste (e-waste). To retrieve metals, computers and electrical wires are being burned, releasing dioxins and PBDEs. In some regions of China where these electronic salvage operations occur, the air and soil contain some of the highest amounts of dioxin measured anywhere (Carroll 2008). These salvage workers and their families are also being exposed to toxic metals associated with this e-waste. In the Chap. 3: 'Environmental Health Curriculum', we describe some lessons focusing on reducing one's risk to e-waste.

These two case studies of common pervasive pollutants—mercury and dioxin demonstrate aspects of exposure, such as: (a) toxicokinetics and toxicodynamics, (b) epidemiology, and (c) the human health consequences. Such an approach, with content adjusted accordingly for the students' knowledge and skills, can be incorporated into a K-12 class to engage students in learning environmental health.

Social and Psychosocial Factors: New Dimensions of Human Health and the Environment

Based on the above discussions, the relationship between environmental health and the biological and physical sciences is quite clear. Environmental health also overlaps with the social sciences. As defined by the WHO, environmental health comprises those aspects of human health that are determined by physical, chemical, biological, social and psychosocial factors in our environment (World Health Organization 2009). Human health is affected by how we interact with and within our natural and built-environments. For example, the mental fatigue associated with urban living and the lack of green space has been shown to be associated with aggressive behaviors (Kuo and Sullivan 2001a, 2001b). In contrast, populations exposed to a green environment exhibit less health inequities (Mitchell and Popham 2008).

While the relationship between the built environment and obesity, a major public health issue, has been recognized (Rundle et al. 2009), it is becoming increasingly evident that neighborhood psychosocial hazards associated with the built environment contributes to disease risk (Augustin et al. 2008), as does interaction with chemical exposures, such as lead (Glass et al. 2009). Psychosocial hazards add a whole new dimension to the discipline and science of environmental health.

Summary

Environmental health is a core discipline of public health. Unlike clinical medicine, public health is aimed at preventing the occurrence of disease rather than treating a disease. In this chapter, some of the basic essentials of environmental health were described and discussed. Several case studies were presented to illustrate how this basic information can be brought together.

Chronic diseases, such as asthma and cancer, occur as the result of the complex interactions between genetics and environmental exposures. The incidences of many of these complex chronic diseases are increasing worldwide. Health disparities occur because of environmental injustices. These health problems and disparities all begin with exposure to hazardous agents and prevention begins by eliminating or reducing these exposures. Many nations have established regulatory agencies to set protective policies, such as removing lead from gasoline. With this policy, blood lead levels dropped significantly in the general US population. Unfortunately, some developing nations often do not have the resources to put policies into place or, as importantly, to enforce them.

Over the last several decades, more and more synthetic compounds have been added to the human environment. As described earlier, we know very little about the hazards that many of these compounds pose. For example, as new technologies, such as nanotechnology, have developed and increased in use, they have been implemented without understanding the potential hazards of nanoparticulates to human health, Nanoparticulates, along with vaporized mercury or diesel fuel exhaust, pose significant dangers because they can pass the blood–brain barrier.

The relationship between the environment and human health is being recognized in places such as Africa (Nweke and Sanders 2009) and the Pacific Rim (Moldren et al. 2008). Understanding that both developed and developing countries are becoming more urbanized, with more inhabitants living in megacities (populations of greater than 10 million), the infrastructure needs and resources required present tremendous challenges for environmental health. Accordingly, the residents of these megacities will face the same, and likely greater, chemical and psychosocial hazards than we are presently observing in urban environments. To meet our present challenges and the new challenges of the future, we need a well-informed and educated citizenry. This can be achieved by incorporating environmental health issues into all components of K-12 education.

References

Allchin, D. (1999). The tragedy and triumph of Minimata: a paradigm for understanding ecological, human-environment, and culture-technology interactions. *The American Biology Teacher, 61*, 413–419.

Augustin, T., Glass, T. A., James, B. D., & Schwartz, B. S. (2008). Neighborhood psychosocial hazards and cardiovascular disease: the Baltimore memory study. *American Journal of Public Health, 98*, 1664–1670.

Barboza, D. (2007). China reportedly urges omitting pollution-death estimates. *New York Times*, C-l.

Brown, V. (2008). Environment becomes heredity. *Environment Magazine*, http://www.miller-mccune.com/…environment/environment-becomes-heredity-4425/

Carroll, C. (2008). High-tech trash. *National Geographic*, pp. 64–87.

Center for the Evaluation of Risks to Human Reproduction (2008). *NTP-CERHR Monograph on the Potential Human Reproductive and Developmental Effects of Bisphenol* A. Washington, D.C.: U.S. Department of Health and Human Services.

Colquhoun, D. R., Goldman, L. R., Cole, R. N., Gucek, M., Mansharamani, M., Witter, F. R., et al. (2009). Global screening of human cord blood proteomes for biomarkers of toxic exposure and effect. *Environment Health Perspectives, 117*(5), 832–838.

Dolinoy, D. C., & Jirtle, R. L. (2008). Environmental epigenomics in human health and disease. *Environmental and Molecular Mutagenesis, 49*, 4–8.

Duncan, D. E. (2006). The pollution within. *National Geographic*. pp.120–134. http://ngm.nationalgeographic.com/2006/10/toxic-people/duncan-text

Dyer, C. (2010). Lancet retracts Wakefields's MMR paper. *British Medical Journal, 340*, 281.

Environment Canada (2009). *Mercury and the Environment*. http://www.ec.gc.ca/mercure-mercury/

Environmental Working Group (2005). Body burden: pollution in newborns. A benchmark investigation of industrial chemicals, pollutants, and pesticides in umbilical cord blood. Accessed 5 August 2009. http://www.ewg.org/reports/bodyburden2

Fraga, M. F., Ballestar, E., Paz, M. F., Ropero, S., Setien, F., Ballestar, M. L., et al. (2005). Epigenetic differences arise during the lifetime of monozygotic twins. *Proceedings of the National Academy of Sciences of the United States of America, 102*, 10604–10609.

Friis, R. H. (2007). *Essentials of Environmental Health*. Sudbury: Jones and Bartlett Publishers.

Glass, T. A., Bandeen-Roche, K., McAtee, M., Bolla, K., Todd, A. C., & Schwartz, B. S. (2009). Neighborhood psychosocial hazards and the association of cumulative lead dose with cognitive function in older adults. *American Journal of Epidemiology, 169*, 683–692.

Goldberg, A. M., & Hartung, T. (2006). Protecting more than animals. *Scientific American, 294*, 84–91.

Grady, D. (2010). U.S. concerned about the risks from a plastic. *New York Times*, A.3

Hargreave, F. E., & Nair, P. (2009). The definition and diagnosis of asthma. *Clinical and Experimental Allergy, 39*, 1652–1658.

Klaassen, E. D. (Ed.). (2007). *Casarett and Doult's toxicology: the basic science of poisons*. New York: McGraw-Hill.

Koch, H. M., & Calafat, A. M., (2009). Human body burdens of chemicals used in plastic manufacture. *Philosophical Transactions of the Royal Society B 364*(1526), 2063–2078. doi: 10.1098/rstb.2008.0208.

Kuo, F. E., & Sullivan, W. E. (2001a). Aggression and violence in the inner city: impacts of environment and mental fatigue. *Environment and Behavior, 33*(4), 543–571.

Kuo, F. E., & Sullivan, W. E. (2001b). Environment and crime in the inner city: Does vegetation reduce crime? *Environment and Behavior, 33*(3), 343–367.

Larmer, B. (2009). The price of gold. *National Geographic*, pp. 34–61.

Legros, G., Havet, L., Bruce, N., & Bonjour, S. (2009). *The energy access situation in developing countries: a review focusing on the least developed countries and subSaharan Africa*. UN Development Organization.

Matchaba-Hove, R. B., Siziya, S., Rusakaniko, S., Kadcnhc, R. M., Dumbu, S., & Chirenda, I. (2001). Mercury poisoning: Prevalence, knowledge and frequency of gold panning and doing retort among alluvial gold panners in Chiweshe and Tafuna communal lands in Zimbabwe. *The Central African Journal of Medicine, 47*(11–12), 251–254.

Matta, I., Villa, L., Ramos, I. M., Sanchez, I., Chopre, G., Ruiz, A., et al. (2003). DNA repair and nonmelanoma skin cancer in Puerto Rican populations. *Journal of American Academy Dermatology, 49*, 433–439.

Mitchell, R., & Popham, F. (2008). Effect of exposure to natural environment on health inequities: An observational population study. *Lancet, 372*, 1655–1660.

Moldren, I., Tao, S., & Carpenter, D. O. (2008). Environment and health in the twenty-first century. *Annals of the New York Academy of Sciences, l140*, 1–21.

Myers, P. & Hessler, W. (2007). Does 'the dose make the poison?': extensive results challenge a core assumption of toxicology. *Environmental Health News.* Retrieved from http://www.environmentalhealthnews.org/sciencebackground/2007/2007-0415nmdrc.html

Nhu, D.D., Kido,T., Naganuma, R., Sawano,N.,Tawara,K., Nishijo, M. et al. (2009). A GIS study of dioxin contamination in a Vietnamese region sprayed with herbicide. *Environmental Health and Preventive Medicine, 14*(6), 353–360.

Northridge, M. E., Yankura, l., Kinney, P. L., Santella, R. M., Shepard, P., Riojas, Y., et al. (1999). Diesel exhaust exposure among adolescents in Harlem: A community-driven study. *American Journal of Public Health, 89*(2), 998–1002.

Nriagu, I. O., Blankson, M. L., & Ocran, K. (1996). Childhood lead poisoning in Africa: A growing public health problem. *The Science of the Total Environment, 181*(2), 93–100.

Nweke, O. E., & Sanders, W. H., III. (2009). Modern environmental health hazards: A public health issue of increasing significance in Africa. *Environmental Health Perspectives, 117*(6), 863–870.

Osburn, W. G., & Kensler, T. W. (2008). Nrf2 signaling: An adaptive response pathway for protection against environmental insults. *Mutation Research, 659*, 31–39.

Ott, W. R., & Robert, J. W. (1998). Everyday exposures to toxic pollutants. *Scientific American, 278*, 86–91.

Porter, S. (2010). Summary of: relationship between mercury levels in blood and urine and complaints of chronic mercury toxicity from amalgam restorations. *British Dental Journal, 208*, 162–163.

Potera, E. (2009). Mass lead poisoning in Dakar: battery recycling exacts a heavy toll. *Environmental Health Perspectives, 117*, A454–A454. doi:1O.1289/ehp.117a454a.

Qian, G. S., Ross, R. K., Yu, M. E., Yuan, J. M., Gao, Y. T., Henderson, B. E., et al. (1994). A follow-up study of urinary biomarkers and liver carrier risk in Shanghuai, People's Republic of China. *Cancer Epidemiololgy Biomarkers and Prevention, 3*, 3–10.

REACH (2007). http://cc.europa.eulenvironmentlchemicals/reachlrcach_intro.htm. Accessed 5 August 2009

Rundle, A., Neckerman, K. M., Freeman, L., Lovasi, G. S., Purciel, M., Quinn, J., et al. (2009). Neighborhood food environment and walkability predict obesity in New York City. *Environmental Health Perspectives, 117*, 442–447.

U.S. Department of Health and Human Services. (2000). *Healthy people 2010: Understanding and improving health,* (2rd ed.). Washington, D.C.: U.S. Government Printing Office

Walsh, B. (2010). The perils of plastic. *Time Magazine, New York*, 42–54

Wei, Q., Matanoski, G. M., Farmer, E. R., Hedayati, M. A., & Grossman, L. (1994). DNA repair and susceptibility to basal cell carcinoma: a case–control study. *American Journal of Epidemiology, 140*, 598–607.

Woolcock, A., & Peat, J. K. (1997). Evidence for the increase in asthma worldwide. In D. Chadwick & G. Cardew (Eds.) *The rising trends in asthma* (pp. 122–134). Wiley: Ciba Foundation

World Health Organization. *Protection of the human environment.* Available on line at hhtp://www.who.int/phe/enl. Accessed on 5 Aug 2009.

Environmental Health Curriculum

Hillary B. Davis, David W. Hursh and Camille A. Martina

Abstract In this chapter, we provide examples of units and lessons on environmental health issues related to air, water, soil and agriculture created by the National Institute of Environmental Health Sciences (NIEHS) funded projects. We provide sufficient depth to enable the reader to implement the lessons suggested in this chapter or to develop their own lessons based on these project lessons. Although literally hundreds, if not thousands of pages of ideas and resources are available on the NIEHS website, we have not attempted to describe all of the lessons. Instead, we provide readers with sufficient examples and guidelines so they can develop and implement lessons on environmental health that best fit the needs of the students and locality. Of course, we do encourage the reader to access the NIEHS and other websites for resources on environmental health information and offer suggestions as to which ones might be appropriate for specific topics. We conclude by describing a model for environmental health curriculum design and implementation.

Keywords Problem-based learning · Air · Food · Water · Soil · Environmental health curriculum · Lessons · Units

Introduction

In the opening chapter we addressed the question, "Why teach students about environmental health?" In this chapter, we examine, "What will engage and educate them? What resources are available? How can these resources be used?"

Students are naturally interested in environmental health because they see the effects in their own environment and lives and realize the need to be proactive in protecting their health. This combination of curiosity, self-interest and the potential to make a difference in their world motivates students to learn core concepts and the

D. W. Hursh and C. A. Martina, *Teaching Environmental Health to Children*, SpringerBriefs in Environmental Science, DOI: 10.1007/978-94-007-1811-1_3,

applications of environmental health within the traditional disciplines. To that end the National Institute of Environmental Health Sciences (NIEHS), in the US, identified the importance of teaching environmental health in the school curriculum.

> The lay community requires greater knowledge about environmental health issues as they are increasingly challenged to make decisions on the risks and benefits of agents that permeate society. In addition, the NIEHS sees these programs as an investment in the future of our society and the environmental health sciences (NIEHS 2006, p. 1).

In other words, all students, even those not pursuing environmental science careers, need to know what the environmental health problems are, and how to gather reliable information to make an informed and preventive decision by examining both the benefits and risks of exposure. These informed decisions are not easily made and offer opportunities for a multidisciplinary approach to problem solving and learning.

Hence, the goal of all nine NIEHS projects was to "improve overall academic performance and enhance students' comprehension of, and interest in, environmental health sciences through integrating EH topics within a variety of subject areas (e.g. geography, history, math, art) and supporting a variety of educational standards." As the Maryland PBS Envirohealth Connections project states, "The environmental health sciences pose complex and engaging questions that can be approached from the perspective of many disciplines. Students appreciate the relevance of the material while teachers can address existing educational standards since students see environmental health topics as relevant to their lives" (Thinkport 2003, 1).

Indeed, the concepts required to understand environmental health issues easily cut across disciplines. Core science concepts are required to understand the physical, chemical and biological processes. Social science topics include environmental health policies and laws both current and historical. Students begin to construct knowledge as to how and why environmental problems emerge within different time periods in response to economic and political interests, or for lack of information. Students could also explore the issue of environmental ethics and justice, and the responsibility to remediate injustices. Mathematics and statistics is needed to understand the scope and effects of environmental health issues. Interpreting and presenting data in graphs and charts, understanding probability, assessing the rate and scope of problems, tracking routes of exposure, evaluating cost and planning mitigation require mathematical skill. English and literature skills are essential to understanding the works of writers such as Carson (1962); Meadows et al. (1977) and Meadows and Wright(2009). Critical reading and research skills are needed to identify reliable sources of nonfiction and integrate information from different sources. Presentation skills are needed to engage with different stakeholders and present findings of local investigations into environmental health issues.

In this chapter, we provide examples from some of the NIEHS funded projects. We hope we have provided sufficient depth to enable the reader to implement the lessons suggested or to develop their own lessons based on these project lessons. Because the different projects together developed literally hundreds, if not thousands of pages of ideas and resources, we have not even attempted to describe all or most of the lessons. Instead, we provide below two charts. The first chart provides an overview of each of the nine projects, including at what institution they were located, project title, grade levels, and website. However, all of the lessons can be found at the National Environmental Health website on *Integrated Curriculum* for teachers: http://www.niehs.nih.gov/health/scied/teachers/curricular.cfm. The second chart is a quick reference guide to the curriculum concepts developed by each project. We conclude this chapter by discussing a model for curriculum design and implementation. The curriculum materials span grades K-12 on issues related to air, water, land and food. Environmental health topics include Table 1 and Table 2:

- Toxicology: germs, disease, household chemicals, solvents, infectious diseases, cancer
- Air: indoor air quality, asthma
- Food: nutrition, contamination, alcohol
- Land/soil: lead poisoning, impact of farming techniques, pesticides
- Water: properties, sources, pollution
- Ethics: environmental justice, societal issues

In the remainder of this chapter our aim is to provide readers with sufficient examples and guidelines so that, even without accessing the websites, they can develop and implement lessons on environmental health.

As indicated above, we begin by focusing on teaching students toxicology, or "the science that studies the harmful effects of our exposure to drugs, environmental contaminants, and naturally occurring substances found in food, water, air, and soil" (AMBIENT 2004a). Consequently, after examining toxicants in and around our home, we provide examples of teaching about soil and water, food borne illnesses, heavy metals (lead), and indoor air pollution.

Toxicology

In the introduction and the chapter on the science of environmental health, we provided information on the science of toxicology. Atmospheric and marine-based interdisciplinary environmental health training (AMBIENT), in expanding on their definition of toxicology in the previous paragraph, states that toxicology includes both the "study of poisons or the adverse effects of chemical and physical agents on living organisms" (AMBIENT 2004b).

The place in which people are typically most often exposed to dangerous chemical toxins or physical agents is the home. Toxins in the home might include household products, which, as we describe below, are the major source for toxic exposures and poisonings. Some exposure sources do not immediately result in poisoning but instead lead to chronic health problems. A few of these toxins and toxicants are found in second and third hand smoke, radon, lead, mold and mildew. Therefore, an ideal starting point for students of any age to begin to understand toxicology is to look at toxins in their home.

Table 1 Overview of the environmental health projects

Project title	Institution	Level	Website
Atmospheric and marine-based interdisciplinary training (AMBIENT)	University of Miami	9–12	http://www.rsmas.miami.edu/groups/niehs/ambient/index.html
ECOS: the environment as a context for opportunities in school	Baylor College of Medicine	K-5	http://www.ccit.bcm.tmc.edu/ceo/content.cfm?menu_id=100
Envirohealth connections: a collaborative exploration of the environment and human health	Maryland Public Tv	6–9	http://www.thinkport.org/default.tp
Project environmental explorations through cross–disciplinary investigation team experiences (EXCITE)	Bowling Green State University	4–10	http://www.bgsu.edu/colleges/edhd/programs/excite/
Hydroville: learning through environmental health science scenarios	Oregon State University	9–12	http://www.hydroville.org/
Integrated Environmental health middle school project (IEHMSP)	University of Washington	6–9	http://hsc.unm.edu/pharnacy/iehms/
My environment, my health, my choices: problem-based learning for environmental health	University of Rochester	5, 7, 9–12	http://www2.envmed.rochester.edu/envmed/ehsc/outreach/index.html
Partnership for environmental education and rural health (PEER): integrating environmental health science in rural schools	Texas A&M University	6–9	http://peer.tamu.edu/IntegratedCur.shtml
Students understanding critical connections between the environment, society, and self (SUC_2ES_2)	University of Medicine and Dentistry NJ	2, 5, 7	http://www.eohsi.rutgers.edu/rc/suc2es2/indexA.shtml

Table 2 Quick reference guide to concepts by project

Project	AMBIENT	ECOS	EXCITE	SUC_2ESS_2	PEER
Grades	7–12	3–5	4–10	2,5, 7	6–9
Child's age	12–18 years	7–10	9–16 y ears	7 years,10,12	11–14
Toxicology—germs, infectious diseases, household chemicals, solvents, cancer	√		√	√	√
Air—indoor air, aasthma, carbon monoxide	√	√	√	√	√
Food—nutrition, contamination, alcohol	√	√	√		
Land/soil—lead poisoning, impact of farming techniques	√		√	√	√
Water—properties, sources, pollution	√	√			√
Ethics—environmental justice, societal issues	√	√	√		√

Project	Envirohealth	Hydroville	IEHMSP	MyEnvironment
Grades	6–9	9–12	6–9	2, 5, 7
Child age	11–14	13–18	11–14	7, 10, 12
Toxicology—germs, infectious diseases, household chemicals, solvents, cancer	√		√	√
Air—indoor air, aasthma, carbon monoxide	√	√	√	√
Food—nutrition, contamination, alcohol	v			√
Land/Soil—lead poisoning, impact of farming techniques	v	√	√	√
Water—properties. sources, pollution	v	√		√
Ethics—environmental justice, societal issues	√		√	√

There are numerous resources on teaching toxicology, including the environmental health projects discussed here. AMBIENT, for example, provides educators with a power point overview "An introduction to the health effects of toxic products in the home" at http://www.asmalldoseof.org/toxicology/home_toxics.php. However, we suggest that teachers and students can together research household toxins using many of the concepts we discussed earlier including dose/response, degree of toxicity, exposure routes, risk, and alternatives.

In the US the National poison data system publishes regular reports on the number of exposure incidences reported and the number of fatalities. Their 2008 Annual Report details that in the US there were over 2.4 million reported exposures, resulting in 1,315 human fatalities that year. The top five substances most frequently involved in all human exposures were analgesics (typically aspirin) (13.3%), cosmetics/personal care product (9.0%), household cleaning substances (8.6%), sedatives/hypnotics/antipsychotics (6.6%), and foreign bodies/toys/miscellaneous (5.2%) (Bronstein et al. 2009, p. 914).

Information on other national poison control centers can be found at the World Health Organization's International Program on chemical safety's website http://www.who.int/ipcs/poisons/centre/directory/en/. The same organization also provides detailed information on ten chemicals or groups of chemicals that are of major public health concern, including: air pollution, arsenic, asbestos, benzene, cadmium, dioxin and dioxin-like substances, inadequate or excessive fluoride, lead, mercury, and highly hazardous pesticides (http://www.who.int/ipcs/features/chemicals_concern/en/index.html).

Because most poisoning events occur in the home (in the US the figure is 90%), having students review dangerous toxins in the home and what they can do to reduce their own and their family's risks can have significant immediate health benefits.

Teachers can work with students to identify some of the toxicants in their home, their degree of toxicity, and exposure routes. For example, the three routes through which humans are exposed to toxicants include ingestion (either directly or, indirectly from hand to mouth contact, as in lead dusk), inhalation, or skin and eye contact. Substances vary in their toxicity. Slightly toxic substances include window cleaners, detergents, and some personal hygiene products and cosmetics, while very toxic substances would include lacquer thinners and solvents, pesticides, ammonia and chlorine products as well as over the counter and prescription drugs. Students can also explore ways in which they can reduce their exposures, either by finding alternatives, developing safe storage methods or following usage guidelines.

Discussion of household toxicants can also lead to in-depth analyses of some of the other toxicants commonly found in the home, such as cosmetics. Some US websites providing information on toxicants in cosmetics include The story of cosmetics: the ugly truth of "*toxins in, toxins out*" at http://storyofstuff.org/cosmetics/, the environmental group website *Not too pretty*, http://www.ewg.org/reports/nottoopretty, and the campaign for safe cosmetics at http://www.safecosmetics.org/). Some of the other topics presented in this chapter are also found in and around the home.

In the US about 400 million electronic appliances are discarded each year, most into landfills. These units often contain heavy metals, such as lead, mercury, and cadmium (Electronic take back coalition 2011). Students can learn about the dangers of these heavy metals and the need for and difficulty of recycling (see *The story of e-waste: why "designed for the dump" is toxic to people and the planet at* http://storyofstuff.org/electronics/). Other sources of toxicants include pesticides and herbicides, and food poisoning.

As students learn how pervasive toxicants in our environment are, they might be interested in knowing more than how to minimize the risk from existing toxicants, but also whether such toxicants should be used at all. The students also learn the advantages of adopting the precautionary principle in which one removes the toxic substance or product and either replaces it with an alternative product or a less toxic substance.

The precautionary principle, Steingraber (1998) writes, states that our primary obligation is to protect human life and, therefore, public and private interests should act to prevent harm *before* it occurs. Consequently, the "indication of harm, rather than proof of harm, should be the trigger for action." (270). Moreover, the precautionary principle requires that we develop the principle of reverse onus, requiring that manufacturers of chemicals and pharmaceuticals demonstrate that what they propose to produce "is almost certainly *not* going to harm anyone" (270, italics in original). Finally, the principle of the *least toxic alternative* "presumes that toxic substances should not be used as long as there is another way of accomplishing the task" (271). The European Commission on the Environment adopted a version of the precautionary principle in February 2000 to "protect the environment and human, animal, and plant health" (European Commission on the Environment 2000). However, other countries, including the US, have failed to adopt similar principles.

Soil and Water

Soil and water compose one of the key pathways through which humans are exposed to toxicants. The primary lesson highlighted in this section was developed and taught in fifth grade (11–12 years old) classrooms. However, the general approach could be adapted for almost any grade level. Another lesson on land, soil, and pesticides is also described in this chapter on teaching about environmental justice and was implemented in a secondary school government class.

Living Downstream: Potable Water and Human Health: A Unit for Late Elementary and Early Secondary Students

In the first lesson described below, students learn how toxicants may unintentionally (for example: oil spills, leaks) or intentionally (for e.g, herbicides, pesticides) come in contact with the soil and ground water. We discuss some of the dangers posed, and what we might do to reduce our risk of exposure. In this particular lesson students researched the dangers posed by herbicides and pesticides. Pesticides and herbicides applied to plants and hard surfaces frequently enter the ground water via irrigation methods or rain. Furthermore, pesticides and herbicide residues that are not washed off can linger or be absorbed into our food supply. Pesticide residues can be found on fruits, vegetables, as well as in the flesh of fish, meat and dairy products. The chronic exposures and ingestion of these residues may pose health problems at very low levels (Steingraber 1998).

This particular lesson was developed and implemented in an elementary school in a suburb outside of Rochester, New York. The history of the community parallels that of many suburban communities that developed in the US after WWII as single-family housing. This community, like so many others, was built on formerly rich and productive agricultural land. Over time, housing and retail shopping is replacing the agricultural land initially spared from development.

In developing the lesson the teachers saw it as an opportunity to introduce students to local history and the environmental land use changes that have and continue to occur. The lesson, therefore, highlights the way that environmental health can complement efforts at place-based education (Smith and Gruenewald 2007; Smith and Sobel 2010). Furthermore, because water plays a central role in the circulation of toxicants, this lesson provides an excellent opportunity to either introduce students to the earth's water cycle or to deepen their understanding of the significance of the water cycle on the health of the planet and its people.

The teachers began with a presentation by the town historian who described the changes in land use. In chronological order and resplendent with photos, maps and artifacts, she described the history of changing land use, including that of the Native Americans, followed first by the European settlers' establishing a vital and prosperous agricultural community from 1860 to 1950, and then by the post-World War II growth of suburbia. The teachers were able to obtain aerial photographs of the part of the community where the school is located to visually document the changing land use from the 1930s to the present. The teachers used the information on the changing land use to ask students to identify when their homes were built and brainstorm possible implications of changing land use on the local environment and their health.

These first historical lessons were followed by lessons that focused on geographic terms and understanding the water cycle. In order to understand how the cycle occurs in their community, students needed to understand geographic water terms such as stream, river, bay, and lake. Students also looked at topographic maps that covered about 25^2 miles around their school. They then used the topographical maps to create three-dimensional modeling-clay relief maps that accurately portrayed how the stream in front of the school flowed down hill into a larger stream, then into a large bay off of lake Ontario, one of the great lakes. The goal was to have students learn how to read topographical maps and understand how the water cycle affects the land and connects bodies of water. The lesson then focused on studying the water cycle so that the students would understand that water evaporated from bodies of water and the soil, condensed into clouds as it cooled, returned to earth as precipitation, and flowed downhill back into streams, lakes, and the ocean. As the water moves over and through the soil, it might also pick-up toxicants such as pesticides and herbicides that affect plant and animal life living in the water. Further, humans catch and eat toxic fish, and cook with, bath in, and drink the water.

The last issue of water quality became the focus on one of the lessons. A teacher handout states:

> Farmland can be the source of sediment, fertilizer, and animal waste pollution. Forests may not be the source of pollutants, but they can be damaged severely by water pollution. Human activities affecting forests (forestry practices such as clear cutting and road construction that can cause erosion and sedimentation) can impact water quality. Cities pose numerous water quality problems due to high water demand, industrial pollutants, non-point source pollution and human wastes.

A student worksheet listed 13 potential sources of water pollution. The student's task was to identify the many ways in which the toxicants from these 13 potential sources might enter the soil and water, as well as to suggest ways to reduce those possibilities. Students were provided with PDF files from the environmental protection agency focusing on water, including "The case of the disappearing water," "Deep subjects: wells and ground water," and "excuse me: is this the way to the drainpipe?" (http://EPA.gov/safewater/kids).

In an effort to help students further understand how our decisions affect the spread of toxicants, including those in water, the next lesson required that students design a fictitious community that included the various components: residential, commercial, recreational, and educational. The students were supplied with a handout depicting a fictitious island that included features such as mountains, rivers, bays, forests, and marshes.

After receiving the handout, the class developed a list of the different elements necessary for most communities, such as: housing, retail stores, industries, airports, streets, highways, harbors, and agricultural and recreational areas. Students also developed their own legends to indicate the different elements on the map. Students worked in pairs to place the different identified elements on the map considering how the different elements might impact the environment and surrounding elements. As they did so, they wrote justifications for their decisions reflecting the environmental impact of the different elements. The assignment required that they think about the relationship between the terrain such as marshes, mountains, plains, harbors, and streams and appropriate land use. Lastly, each student presented and compared their maps with others in the class.

In the next lesson students identified the chemicals found within common household products and researched the effects of those chemicals on human health. This gave students an opportunity to search the environmental protection agency website (http://www.epa.gov/kidshometour) and record their findings so that they could use the information for the next activity.

For the culminating activity for this lesson, students were directed to research pesticides and herbicides for their impact on human health and their community. In addition, the students initiated on their own researching the environmental impact of pet waste, which if not properly disposed of, poses significant health problems. After researching the environmental impact, they researched possible solutions to reduce or eliminate the problem at the individual and community level, and presented what they had learned to other students and school community through websites, brochures, power point presentations, and videos. Among the websites students used included http://www.epa.gov/kids/.

Since students had input on what they wanted to research and how they would present it to their classmates, they took more initiative than usual. Teachers observed that students found the project to be more motivating than their traditional classroom work.

We depict a similar project on toxicants in the environment in the chapter on working for environmental and social justice. In that chapter, we describe a unit taught by Michael Fantauzzo in a course at the secondary level on civics, in which students investigated five different potential environmental health problems: (1) pesticide use, (2) lead poisoning, (3) electronic waste, (4) governmental recycling bottle programs, and (5) hazardous materials transportation. Rather than repeating the entire description here, we refer you to the description in the next chapter. Here, we note that the example focusing on pesticides and herbicides that begins with a fictional letter, written by the fictional Ivana Kleenup, is a good illustration of students becoming engaged in evaluating the effects of chemicals on our food and in our drinking water.

Food

In the US, the past several years have brought frequent food safety warnings and outbreaks of resulting illness and death from bacteria such as *salmonella* in eggs, cantaloupe, peanuts, peanut butter, jalapeno peppers, and tomatoes as well as *E. coli* found on prewashed spinach or in processed beef products. The sources of such contaminations are often difficult to pinpoint because large scale food producers distribute products to many manufacturers and companies. One such example occurred in January 2009 when eight people died and 500 became ill from *salmonella* contaminated peanut butter. After the outbreak only 22.5% of consumers in a study by the University of Minnesota's Food Industry Center (2009) were confident that the food supply was safer than a year earlier. In a study by the Consumer Reports National Research Center in January 2009, 83% of consumers were concerned that harmful bacteria or chemicals are contaminating their food.

In this section, we describe activities about food borne illnesses. The first set of activities has been used with high school students in Project AMBIENT to influence their personal food preparation, handling and storage behaviors at home and their awareness of food safety when buying or consuming food outside the home. In the Project ECOS My World, students in grades 3–5 learn about food sources, healthy eating, safe food preparation and handling, the basic food groups and overall nutrition through a storybook, activities and a mini-magazine. In the Envirohealth project, students in grades 6–10 explore food borne illnesses, and conduct experiments to understand the concept of anti-oxidants, and to learn about genetically modified foods.

Atmospheric and Marine-Based Interdisciplinary Environmental Health Training

In the AMBIENT module on food borne illnesses, students begin with a PBL simulation in which they are on an ocean cruise. They order from a menu and some of them get sick. To figure out why only some people got sick, they analyze the menu for possible sources of food poisoning. Since this scenario is based on actual outbreaks on board cruise ships, students watch video clips from news stories and ask questions about how food borne illnesses occur. The videos can be accessed at: http://www.rsmas.miami.edu/groups/niehs/ambient/teacher/food/Tfood.html.

Students are further enticed into food preparation issues by a teacher created video. This hilarious spoof shows a naïve customer ordering from a menu that includes "Sal Moan Ella" and a chef who uses the same knife and cutting board to prepare raw meat and to cut up vegetables for a salad. The challenge for students is to identify all the dangerous practices in the video. Next, the students use an actual health department inspection form hazard analysis and critical control point system (HACCP) to look at their own home kitchens and a commercial kitchen such as the school kitchen. This highly detailed checklist includes a physical inspection of the food preparation area, the preparation practices of the people, food storage in the refrigerators and freezers, utensil washing and storage, food sources and transportation, general food storage and serving and eating areas. Students realize quickly that even small changes can help prevent food contamination. Educators can obtain or download health inspection forms to use for this exercise.

In food safety labs, students use yeast cultures to simulate how high temperatures affect bacterial growth, how low temperatures affect food preservation, how cross-contamination of bacteria can occur, and how poor hygiene affects bacterial growth. For example, in the high temperature lab, they see the exponential growth and decay of the microbial populations in relation to the organism's temperature tolerance range and how high temperature inhibits microbial growth. In the low temperature lab, students set up an experiment comparing the decay of fruit kept in the refrigerator versus fruit left out. Students take temperature and humidity readings every day at the same time, and then track the growth of the bacteria when it appears. Students keep careful notes in their lab books about their experiment, which includes their sketches and daily data. They use these sources and evidence to support their conclusions about the effect of temperature and humidity on bacterial growth.

To further develop student's understanding of how food can become contaminated in unintentional ways, students document their own family and cultural food practices. In applying what they have learned, they reconsider the safety of food preparation around holidays. For example, often with the distractions and socializing during a holiday party, food may be left out at room temperature much longer than the 90 min safety window. This is also the time when there are multiple cooks in the kitchen, and utensils and surfaces are used for different purposes. What could be potential problems in this holiday scenario? What unintentional health consequences could evolve out of misinformation about food preparation? These are identified in the food myths (in the appendix) exercise where students identify fact versus opinion statements about food safety. Through sharing their own food myths and further reading, they draw conclusions about how culture, geography and climate affect what people eat and the healthy and harmful food practices that arise from them.

With their new understanding of food safety, students turn to investigate a food borne illness outbreak. Using a more detailed cruise ship scenario, students examine specific data from a random selection of reported cases of illness. Students use a symptom tally worksheet (available on the website as foodborne illness outbreak investigation: <http://www.rsmas.miami.edu/groups/niehs/ambient/teacher/food/Tfood.html> to determine the most commonly occurring symptoms. Then they graph the epidemic curve showing the number of people ill versus the date of onset of the symptoms. Based on interview data from a random selection of banquet attendees, students use an attack rate worksheet to calculate the attack rate for each food. Using these data, they determine the suspect foods and hypothesize how they became infected with data from tables of shellfish related infectious disease (viral, bacterial and parasitic) and chemical (natural marine toxins made by dinoflagellates). Students plan how they might confirm their diagnoses, such as inspection of shellfish and the shellfish bed, or testing for biomarkers of the disease in the humans.

The next set of labs helps students understand more about dinoflagellates, the phytoplankton at the base of the marine food web that produce potent natural toxins that accumulate in the seafood we eat. Humans exposed to these dinoflagellate toxins through contaminated seafood result in both acute and chronic health effects. So scientists need to identify species types, levels of activity and concentration of dinoflagellates. Students do labs on chlorophyll, photosynthetic pigments and bioluminescence where they use spectroscopy and chromatography used by scientists in monitoring dinoflagellates.

To integrate all the knowledge developed in the AMBIENT food module, students develop a pamphlet, slide show, or other educational material to teach a specific population (such as younger children or grandmothers) about preventing food borne illnesses. Students are prompted to create an eye-catching, engaging oral and written presentation that will influence people with factual information. It is especially important that students come to see themselves as educators.

An extension activity on food safety has students research the safety of genetically modified foods and debate the pros and cons. This requires students to apply what they have learned about food and food safety to understanding the nature of GMOs, the conditions of growth, nutritional value and potential health risks.

Lead: A Heavy Metal

Heavy metals, including mercury, zinc, arsenic, selenium, and cadmium, often pose a danger to our health. One of the more likely heavy metals with which we are liable to come into contact via water, soil, and air contamination is lead. Lead, can be found in pre 1978 house paint in the US, toys, jewelry, and other manufactured goods. Fortunately, there are actions people can take to reduce their exposure. Therefore, understanding the toxic properties of lead, our exposure risks and how we might reduce them can be an empowering lesson for students. In the following section, we begin by describing some of the likely routes for lead poisoning, its effects, and preventative strategies to reduce their risk to exposure.

Lead Poisoning

In the US, the most likely source of lead poisoning comes from lead in lead-based paint. While France, Belgium, and Austria banned lead-based interior paint in 1909 and the League of Nations followed in 1922 (Kitman 2000), lead-based paint continued to be used in the US until 1978. However, recent analyses of enamel paint sold in twelve countries in Asia, Africa, and South America revealed that in all of the countries, the paint had average concentrations of lead that far exceeded (by multiples of 10–500) even the highest national limits placed lead (Scott-Thomas 2009). Scott-Thomas (2009) cites research by Fewtrell et al. (2004) who estimate that 20% of all children have blood lead levels above the maximum allowable of 10 mg/dl and that most of these children live in developing countries.

One of the ways in which people and especially children are exposed to lead in the home is from lead paint dust that is generated by the opening and closing of windows and from poorly maintained walls that through decay shed painted plaster or wallboard onto the floor. Young children, then, are exposed to lead as they play or crawl on the floor and ingest lead when they put their fingers in their mouth, inhale dust into their lungs, or penetrates their skin.

Children and adults can also be exposed to lead through water that has been sitting for more than 6 h in lead pipes (found primarily in older homes), through toys and other objects painted with lead based paint, and eating food harvested from soil that contains lead left over from auto exhaust before lead was removed from gasoline (petrol).

Persons living in developing countries face additional exposures to lead. As stated above, while lead has been banned from paint and gasoline in some countries (most notably western European countries) lead-based paint and gasoline is still manufactured and used in some Eastern European, the Middle East, Northern Africa, Southeast Asian, Latin American, and African countries (http://lead.org,au/fs/fst27.html). Students could investigate whether lead-based gasoline is banned in their own country, what is the allowable limit on lead in paint in their country and whether that limit is enforced, and what can be done to strengthen and enforce the regulations.

Another burgeoning lead risk in developing countries comes from the danger of lead poisoning from car batteries where they are melted down and sold as scrap metal (Gottesfeld 2009). A horrific example reported by the World Health Organization (WHO 2008, 2009) was the exposure to lead that had been recovered from batteries that resulted in the poisoning of several hundred residents and 17 fatalities in Thiaroye Sue Mer, Senegal. Perry Gottesfeld (2009), executive director of occupation knowledge international reports that "120 million people are overexposed to lead around the world. That is three times the number affected by HIV AIDS."

At What Level is Lead Toxic and What are the Consequences?

Lead, like other heavy metals, is biopersistent, which means that once they are inside a living organism, they remain there for a really long time. Furthermore, lead, like mercury, bioaccumulates and collects in the bones and brain.

How much lead people have in their bodies is measured by the amount of lead in the blood, and is dangerous at even the lowest levels. While the Centers for Disease Control (CDC) states that there is no safe level of lead in the blood, in the US a level of >10 μg of lead/dl of blood (10 μg/dl) is sufficient for the CDC to recommend that public health actions be initiated. It is estimated that 250,000 children in the US age 1–5 years have lead in their blood to that degree (CDC http://www.cdc.gov/lead/). Such levels have been linked to learning disabilities and reproductive disorders (Schettler 2003).

At higher levels (50–100 μg/dl), many organs can be affected. Young children, younger than 6 years old, are especially susceptible to harm from lead poisoning. They can suffer neuronal (brain cell) damage, which is not reversible. Lead poisoning can also replace bone marrow and, therefore, disrupt hemoglobin synthesis and the making of red blood cells, resulting in anemia. More lead can damage kidney cells, which may lead to kidney failure, can be stored in bones affecting bone formation, can negatively affect the peripheral nervous system, and can lead to numerous symptoms, including headaches, abdominal pain, vomiting, anemia, weight loss, learning difficulties, hyperactivity, seizures, coma, and even death (CDC http://www.cdc.gov/lead/). In some cities in the US, as many as one out of four children are thought to be lead poisoned. A lead awareness program is available in the US through the environmental protection agency (http://www.epa.gov/lead).

Children are more susceptible to toxicants than adults because they are smaller in size and have a higher metabolism. Environmental health (2004) suggests a lesson to demonstrate to students how lead affects children and adults differently. Two clear containers of different sizes (the larger should hold about ten times more liquid than the smaller) are used to represent the amount of blood in an adult and a child. Then, the same amount of food coloring (10 mcg/dl is suggested) is dropped into each container. The smaller container, representing a child, will have the darker water, demonstrating that with the same amount of lead, the concentration of lead in the child is greater.

Before proceeding to describing some additional lessons, in the US childhood lead poisoning is often considered an environmental justice issue. Children who live in homes built before lead paint was banned in 1978 are at highest risk of being lead poisoned. These homes are often dilapidated and in high poverty urban and rural areas in which the children are the often the individuals who suffer the most.

In some counties, states, and countries, students should be able to obtain the concentrations of lead in blood of children by income and ethnicity. In the US, black and hispanic (children ages 1–5) have higher median concentrations of blood than white children and the lower the income of a child's family, the higher their lead concentration is likely to be (Environmental Protection Agency, 2010).

Suggested Lessons on Lead

All of the above suggests that teaching about lead and lead poisoning can contribute significantly to the health of children and their families. It also provides an opportunity to demonstrate to students that they can act to improve their health.

The interdisciplinary environmental health curriculum for middle school students (IEHMSP, http://hsc/unm.edu/pharmacy/iehms) curriculum has several different units related to lead and lead poisoning, most of which can be easily adapted for a specific school and community. Since this particular curriculum resource is 349 pages long, we will only summarize a few here.

1. The first lesson provides students with the background knowledge they need on environmental health and lead. It begins by defining environmental health at "the study of how the environment affects human health" (p. 7) and gives examples of things that may affect your health. Next, the lesson defines toxicity exposure, dose/response, and individual susceptibility. In addition, they too, provide information on how children might come into contact with lead, its long-term health effects, and what can be done to reduce your risk. With a wet cloth, regularly wipe the dust from toys, counter tops, tables, and window sills and mop the floor. Because water that sits in lead pipes overnight or for long periods of time absorbs lead, pipes should be flushed for 30 s before the water is used for drinking or cooking. Lastly, people need to be aware of other potential sources of lead poisoning, such as when removing old paint from furniture or houses.
2. The second lesson address lead poisoning during the Roman empire. Students are provided information about the many ways that those living during the ancient Roman empire came into contact with lead. These include the lead-lined Roman baths in Bath, England and elsewhere, wine sweetened with a

sweet syrup made from grapes in a lead pot or lead-line copper kettle, and aqueducts that had lead-lined pipes (p. 37). Students also learn that Pliny the (23–79 AD) noticed that because the wealthy were also more likely to enjoy wine sweetened with sappa, water carried to their community via aqueducts, and use lead utensils, that wealthier Romans were more likely to fall victim to diseases such as colic. Finally, the students create a travel guide for a time travel vacation to the Roman empire giving suggested health precautions.

3. The third lesson is on the geography of childhood lead exposure. Because those who suffer from lead poisoning in industrialized countries like the US are more likely to be living in poverty or racial minorities, and globally, the children more likely to suffer live in developing countries, lead poisoning is a social justice issue. As described above, in the US, students can access information on instances of lead exposure by geographical areas. By mapping where such instances occur and knowing something about the areas, students can analyze the environmental and economic conditions that contribute to lead poisoning.
4. In the fourth lesson both students and teachers work together to research the likely sources of lead poisoning in their communities. They explore which population demographics are more likely to be afflicted, and what can be accomplished to reduce children's exposure through promoting new regulations, as well as the enforcement of existing regulations, and/or changing personal behaviors (such as frequently damp mopping floors on which young children might crawl and play).
5. The fifth lesson addresses cultural beliefs and 'cures' that result in toxic lead exposures. Around the world, lead is often used in folk remedies and candles, thus exposing their users to potential lead poisoning. In the IEHMSP curriculum, students read case studies about Hispanic children who were exposed to lead via these routes. Teachers, working with students, can research what is known about lead in folk remedies, candles, and other commodities in their community. In the US, lead is sometimes found in cosmetics, particularly in lipstick and eyeliner (http://www.safecosmetics.org). While the lead levels in cosmetics may be low, the Centers for Disease Control and prevention state that there is no safe level of lead in blood.
6. In the lesson "Eighteen pence a day" students examine "the working conditions in London's white lead mills in the mid-1800s by reading biographical sketches of Charles Dickens and Jack London. Then, they read excerpts from Dickens' *The Uncommerical Traveller,* and conclude reading Jack London's description of a young factory worker in *The People of the Abyss* and write a journal entry for the worker's point of view.

Children and adults in developed and developing countries can take many actions to reduce their exposure to lead. In the US, the environmental protection agency (http://www.epa.gov/lead/pubs/leadinfo.htm#protect) has numerous suggestions for how to reduce one risk of lead poisoning. The same guidelines can be applicable for most situations but leaves out additional risks found in developing countries, such as the risk of exposure from recycled car batteries, computers, and other electronic waste.

Air Quality: Problem-Based-Learning

Inhalation of contaminated air is also a key pathway through which humans are exposed to toxicants. Just as potable water is essential for life so is the air we breathe. The unit on air quality we highlight in this section was designed for a secondary school level environmental science class. The central curriculum design of this unit is problem-based-learning (PBL). In PBLs students learn the skills to approach a complex problem through analysis of: (1) *What they know and* (2) *What they need to know* through the development of a problem statement. PBL is a student-centered instructional strategy in which students collaboratively solve problems and reflect on their experiences. Characteristics of PBL are that learning is: (1) driven by challenging, open-ended problems, (2) small collaborative groups work to solve the problem and (3) teachers are facilitators of learning. In addition, students take responsibility for their collaborative group and organize and direct the learning process with the support of their teacher. PBL can be used to improve content knowledge and promote and refine the development of communication, problem-solving, and self-directed learning skill. The problem presented in this unit can be used in a wide range of classes including health, biology, ethics, language arts and social studies.

When Dan Sullivan, a city high school biology teacher in Rochester, New York inquired what environmental health issue his students' wanted to explore, his students responded to the higher indices of asthma amongst children who live in urban areas (Malveaux and Fletcher-Vincent 1995) In meeting his students' concerns, Sullivan developed a 14-session PBL unit *Asthma in the city.* Free access to the complete14 session curriculum and materials of *Asthma and the City* can be found at http://www2.envmed.rochester.edu/envmed/ehsc/outreach/index.html.

In the first session, the students are given copies of *separated at birth*, a fictional description of adolescent identical twins who were separated at infancy and then reunite only to learn that while they have identical genes, the twin who grew up in poverty in the city has asthma while the twin who grew up economically advantaged does not. This text of the handout is simple, short story that teachers could create to fit the social and economic context for their situation. Sullivan's example fits the US context where those living in poverty often live in substandard housing in urban areas.

The students in the class, therefore, are asked to explain why one twin has asthma while the other twin does not. To answer this quandary, the next 13 sessions facilitates the students learning with more information being revealed over time, as well as the students' own investigative work. Some of the information given is intentionally misleading or "red herrings". The use of red herrings helps students through their own investigative process, queries and

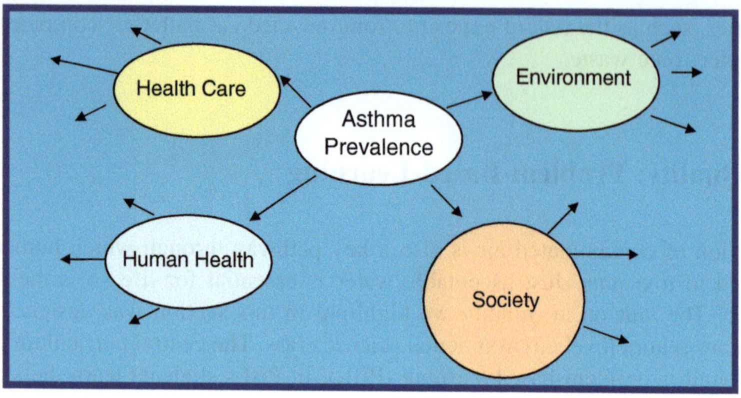

Fig. 1 Sample categorical web framework

collaborative discussions learn how to discern information as to what is important and/or credible evidence and what is neither credible nor useful. In the growing field of environmental health there is much information, some credible and some myth. We want students to be able to discern the difference.

The students begin by forming collaborative learning groups. These heterogeneous groups composed of four students each are given the title of a *Think-Tank*. Each *Think-Tank* investigates the problem by assuming four different professional roles. The roles are that of: (1) a medical doctor (health care), (2) environmental biologist (environment), (3) epidemiologist (human health) and (4) sociologist (society). Poster paper is provided to each group and they are asked to brainstorm ideas about the scenario. "What is happening? What is the problem?" "Can the environment have an impact on resulting health problems?" The teacher facilitates the class discussion and through careful questioning asks general questions that do not divulge essential information. At this point the class creates a categorical web that will remain posted in the classroom throughout the next 13 class sessions. The web should focus on four areas of the problem that will be linked to the *Think-Tank* roles: an example of what this categorical web might look like (see Fig. 1).

(1) A medical doctor (health care), (2) environmental biologist (environment), (3) epidemiologist (human health) and (4) sociologist (society) (adaption from *my environment, my health, my choices,* 2005, *reprinted with permission*).

The teacher then asks the *Think-Tanks* to pose at least four questions they have about the problem and write theses question on paper to hand in at the end of class. The students are also informed that will receive more information about the problem at the beginning of the next class, with the caveat that this information may or may not provide the students with the information they need to answer their questions.

In session one, the teacher is encouraged to review question lists from each group and to use these questions for a *question and answer update* that may reveal

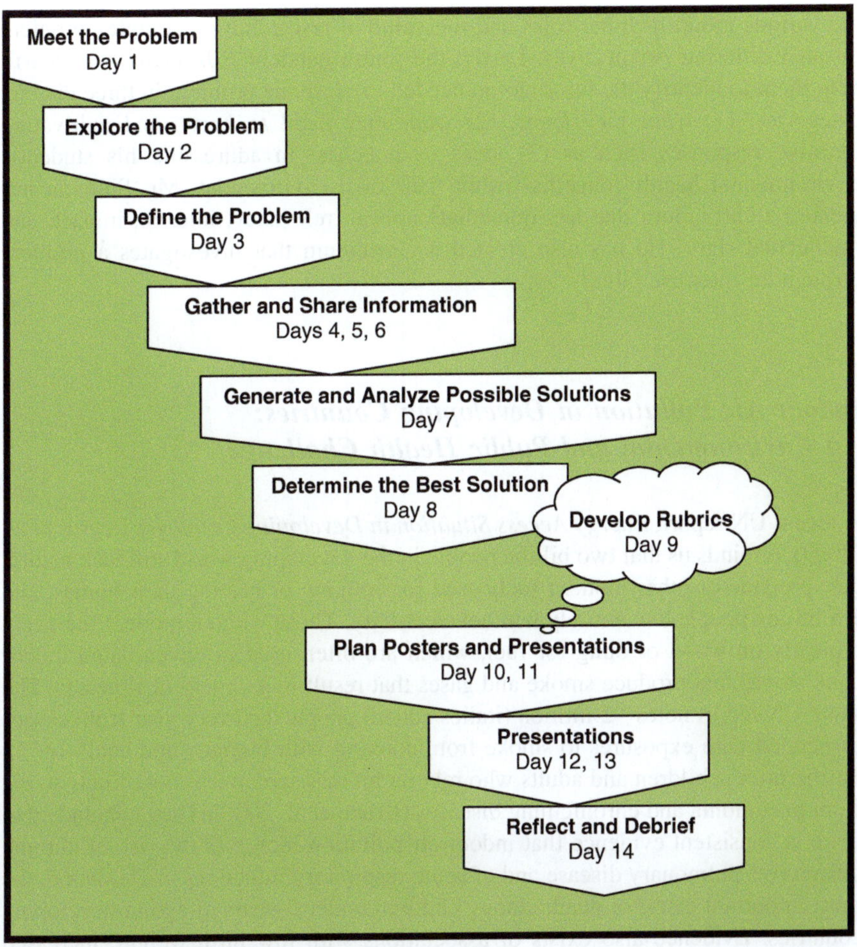

Fig. 2 Problem-based learning timeline

possible answers to the questions. The goal of the daily update is to motivate students to become active in the process of learning and to document and make evident this process over time.

In the second session, *day two: preparing to solve the problem*, each student is given the following four materials. the first is the *PBL timeline map* that explicitly explains steps involved in solving a problem per curriculum unit day and title (see Fig. 2, *problem-based learning timeline,* my environment, my health, my choices © 2005, University of Rochester, reprinted with permisson).

The second handout is the *collaborative work rubric* to review cooperative-learning expectation*s*. The third handout describes separately the four roles of the *Think Tank* professions: (1) medical doctor, (2) epidemiologist, (3) environmental biologist and (4) the sociologist. These *role cards* identify, describe and explain the various group member roles and the value of researching a complex problem through differing perspectives. Lastly, the fourth handout *PBL chart*, which will help them to identify the information needed to solve the problem in three distinct categories; (1) *what they know*, (2) *what they need to know*, and cultivating intuitive responses such as (3) ideas or hunches. In addressing his students' environmental health concerns within their own environment, Mr. Sullivan has created a curriculum that has immediate appeal, relevancy, interest, impact, and intellectual rigor. He has also created a curriculum that investigates a problem through an interdisciplinary lens.

Indoor Air Pollution in Developing Countries: An Environmental and Public Health Challenge

A recent UN report, *Energy Access Situation in Developing Countries* Legros et al. (2009), reminds us that two billion people in the developing world still lack natural gas, propane or other modern fuels used for cooking or heating their homes, and 1.5 billion people live entirely without electricity. Those without modern fuels rely typically on wood or dung for fuel, which are often used in unventilated indoor cook stoves that produce smoke and gases that result in a variety of illnesses. The same UN report notes "2 million (italics added) people die every year from causes associated with exposures to smoke from cooking with biomass and coal" (p. 2). Furthermore, children and adults who rely on biomass are much more likely to die from pneumonia and chronic lung diseases (Bruce et al. 2000). They conclude that there is consistent evidence that indoor air pollution increases the risk of chronic obstructive pulmonary disease and of acute respiratory infections in childhood, the most important cause of death among children under 5 years of age in developing countries. Evidence also exists of associations with low birth weight, increased infant and perinatal mortality, pulmonary tuberculosis, nasopharyngeal and laryngeal cancer, cataracts, and, specifically in respect of the use of coal, with lung cancer (1078).

Rather than an imaginary scenario, students in developing countries could engage in real PBL. They could gather and examine evidence that cooking with biomass on open fires or poorly function stoves results in numerous health problems. Further, they could research the technological, cultural and economic barriers to adopting remedies. For example, some people are beginning to use rocket stoves that are more efficient but may not know why they are better or how to properly use them (see Bilger 2009). Moreover, culture sometimes plays a role as some communities have cooked with particular fuel sources for generations and wood fires or barbecuing in the US is preferred because of wood or charcoal imparts on the food. Lastly, the development of more sustainable, reliable, and healthier fuel sources depends in part on the socioeconomic developments in developing countries.

Students can play a significant role in documenting environmental health problems and educating the community in the relationship between the cooking with biomass over open fires and health problems, and help the community work towards solutions. Hursh (2010) describes his recent efforts in Uganda working with a school to teach students about energy and install a rainwater harvesting system so that the school has water for showers and toilets.

Interdisciplinary Lens of Environmental Health

We have adopted an interdisciplinary approach (1) to facilitate and support the development of the students' ability to examine an environmental health problem through a variety of disciplines and ways of knowing and (2) to understand the importance of air, water, soil, and other vectors on health. In order to meet these objectives the students will be encouraged to gather, organize, and synthesize information from various sources, and identify potential ways in which to address a problem. This approach challenges the student to not only explore the problem, but to also research and answer the question in a variety of ways, through varying perspectives. This will occur through the dynamic and dialectical processes of collaborative learning. They will also have to discover ways to evaluate their new knowledge and to defend their proposed answers to other class members.

Learning about environmental health can and should be integrated into the curriculum of both elementary and secondary schools. Moreover, what students learn about the dangers of different toxicants to their health and what they can do to reduce their risk can be crucial to improving their own health and the health of those around them. Schools and students can become leaders in creating a safer and socially just environment, which is a topic we elaborate on in the next chapter.

References

AMBIENT (2004a, January 13). Toxicology—Relating to environment health. Coral Gables FL: University of Miami

AMBIENT (2004b, April 1). Toxicology and you: Chapter 1. Slide show developed by AMBIENT.

Bilger, B. (2009, December 21). Hearth surgery: The quest for a stove that can save the world. *The New Yorker, 85*(42), 84.

Bronstein, A. C., Spyker, D. A., Cantilena, L. R., Jr., Green, J. L., Rumack, B. H., & Giffin, S. L. (2009). 2008 Annual report of the American Association of Poison Control Centers' National Poison Data System (NPDS): 26th Annual report. *Clinical Toxicology, 47*, 911–1084.

Bruce, N., Perez-Padilla, R., & Albalak, R. (2000). Indoor air pollution in developing countries: A major environmental and public health challenge. *Bulletin of the World Health Organ, 9*(78), 1087–1092.

Canfield, R. L., Jr. Henderson, C. R., Cory-Slechta, D. A., Cox, C., Jusko, T. A., & Lanphear, B. P. (2003). Intellectual impairment of children with blood lead concentrations below 10 μg per deciliter. *New England Journal of Medicine, 16*(348), 1517–1526.

Carson, R. (1962). *Silent spring*. New York: Houghton Mifflin.

Clark, C. S., Rampal, K. G., Thuppil, V., Roda, S. M., Succop, P., Menrath, W. et al. (2009). Lead levels in new enamel household paints from Asia, Africa and South America. *Environmental Research, 109*, 930–936.

Electronic Take Back Coalition (2011). E-Waste problem overview. Accessed at http://www.electronicstakeback.com/resources/problem-overview/

Environmental Protection Agency (2003). *America's children and the environment.* http://www.epa.gov/economics/children/body_burdens/b2-graph.html and http://www.epa.gov/environhealth/children/ace_2003.pdf

European Commission on the Environment (2000, February 2). Commission adopts communication on precautionary principle. Brussels: European Commission on the Environment.

Fewtrell, L., Pruss-Ustun, A., Landrigan, P., & Ayuso-Mateos, J. (2004). Estimating the global burden of disease of mild mental retardation and cardiovascular disease from environmental lead exposure. *Environmental Research, 94*(2), 120–133. February.

Gottesfeld, P. (2009, February 20). Lead in the developing world. *Living on Earth* Accessed December 5, 2010. http://www.loe.org/shows/segments.htm?programID=09-P13-00008&segmentID=8

Hursh, D. (2010, August). Education for global sustainability: A case study of a school in Uganda. Unpublished manuscript.

Kitman, J.L. (2000, March 20). The secret history of lead. *The Nation.* 11–40. Accessed December 5, 2010. http://www.thenation.com/article/secret-history-lead?page=0,2

Legros, G., Havet, I., Bruce, N. & Bonjour, S. (2009). The energy access situation in developing countries: A review focusing on the least developed countries and Sub-Saharan Africa. UNDP-WHO. Available at http://content.undp.org/go/newsroom/publications/environment-energy/www-ee-library/sustainable-energy/undp-who-report-on-energy-access-in-developing-countries-review-of-ldcs—ssas.en

Malveaux, F. J., & Fletcher-Vincent, S. A. (1995). Environmental risk factors of childhood asthma in urban centers. *Environmental Health Perspectives, 103*(6), 59–62.

Meadows, D. H., Meadows, D. L., Randers, J., & Behrens, W. W., III (1977). *The limits to growth A Report for the club of Rome's Project on the predicament of mankind.* New York: New American Library.

Meadows, D. H., & Wright, D. (2009). *Thinking in systems: A primer*. London: Earthscan.

NIEHS Center for Ecogenetics & Environmental Health at the University of Washington, Seattle and the Niehs New Mexico Center for Environmental Health Science. (2004). Lead Integrated Environmental Health Middle School Project.

NIEHS: environmental health science as an integrative context for learning (2006). *Accomplishments, impacts and next steps: planning for tomorrow.* Annual grant meeting, June 22–23.

Schettler, T. (2003, February 8). Developmental disabilities—impairment of children's brain development and function: The role of environmental factors. *The collaborative on health and the environment.* Accessed at http://www.healthandenvironment.org/learning_behavior/peer_reviewed Paper adapted from: Schettler T. Toxic threats to neurologic development of children. *Environmental Health Perspect.* 2001 Dec;109 (Suppl 6:813–6).

Scott-Thomas, C. (2009). *Survey shows consumer concern about preparation, not supply.* Retrieved January 10, 2010. http://www.foodnavigator-usa.com/Financial-Industry/Survey-showsconsumer-concern-about-preparation-not-supply. Retrieved 10 January, 2010.

Smith, G. A., & Gruenewald, D. A. (2007). *Place-based education in a global age: Local diversity*. NY: Routledge.

Smith, G. A., & Sobel, D. (2010). *Place- and community-based education in schools*. NY: Routledge.

Steingraber, S. (1998). *Living downstream: A scientist's personal investigation of cancer and the environment.* NY: Vintage Books.

University of Minnesota (2009, February 24). Confidence in food safety plunges in wake of peanut butter contamination. *Science Daily*. Retrieved December 27, 2010. World Health Organization (2008, June 23). Lead intoxication in Senegal. World Health Organization: *Global Alert and Response*. http://www.sciencedaily.com/releases/2009/02/090224133018.html, http://www.who.int/csr/don/2008_06_23/en/index.html

Working for Social and Environmental Justice Through Environmental Health

David W. Hursh, Camille A. Martina and Hillary B. Davis

Abstract Natural and synthetic toxicants can cause harm to people and their environment. People who are living in poverty typically face greater environmental risks than people who are wealthy. Consequently, we can ask whether these differing health risks are fair or just, and what we, as teachers and students, can and should do about it. We can also ask how do we create a society that will benefit both people and the environment. In this chapter, we show how these questions can lead to students changing their behaviors and working to educate their fellow students and community. Students can also act to change environmental policies at the local, state/provincial, national, and global levels in order to reduce or eliminate exposure risks. In the process, students can gain a better understanding of what they can do as citizens to create a society that promotes the health of all its citizens.

Keywords Human rights · Environmental justice · Ecojustice · Place-based education

Introduction

Natural and synthetic toxicants in the environment often harm people and the world around them. Moreover, people who are less powerful typically face greater environmental risks than those who are wealthier or in the dominant racial group (Bullard 1993, 1994). Consequently, we can ask whether these differing health risks are fair or just, and what we, as teachers and students, can and should do about it. In addition, we can ask how do we create a society that will benefit both people and the environment.

In this chapter, we show how these questions can lead to students changing their behaviors and working to educate those around them. Students can also act to

D. W. Hursh and C. A. Martina, *Teaching Environmental Health to Children*, SpringerBriefs in Environmental Science, DOI: 10.1007/978-94-007-1811-1_4,

change environmental policies at the local, state/provincial and national levels in order to reduce or eliminate exposure risks. In the process, students can gain a better understanding of what they can do as citizens to create a society that promotes the health of all its citizens.

We begin this chapter with an overview of how the notions of social and environmental justice have evolved since the 1970s. Rather than seeing the evolving conceptions of justice as a problem because there is not one agreed upon definition, we see it as an opportunity to explore a range of questions over what justice means and how it can be achieved.

In the second half of the chapter we provide in-depth descriptions of lessons and units in which students examined environmental health within the context of environmental justice. We first describe how fifth graders (11–12 year olds) researched the dangers from pesticides to herbicides, explored what they can do to reduce their own exposure, and then developed different media presentations to educate students, teachers, parents, and community members. We also describe students in a culminating secondary civics class who researched the problems associated with pesticide use, lead poisoning, electronic waste, and hazardous materials transport and then developed public policies to reduce or eliminate the dangers. Next, we turn to an overview of the ethics module at AMBIENT (2004), the program at the University of Miami. In this module, secondary students are educated regarding the key concepts in ethical decision making as they explore case studies examining toxicants in the air, water, soil, and food. In all of these lessons, students learned to use ethical reasoning protocols to create productive and civil discourse in urban high schools.

What is Social Justice, Environmental Justice, and Ecological Justice?

Throughout this book we have been building on the National Institute of Environmental Health Sciences definition of environmental health as the field of science that studies how the environment, which includes both the natural and the human created environment, "influences health and disease" (NIEHS 2005, p. 1). Implicit in that definition is the reciprocal relationship between people and their environment. Humans affect the environment and that environment positively and negatively affects human health.

However, questions of what *should* be the relationship between humans and their environment and the subsequent unequal effects that the environment then has on humans pushes us beyond a scientific understanding of the relationship between the environment and health, to questions of ethics, governance, history, and policy.

Furthermore, we should note from the beginning that there are no agreed upon definitions of social justice, environmental justice, and ecological justice, and reasonable people differ regarding which conceptions should be adopted and promoted. We suggest that it is just these uncertain aspects of environmental health that make it a rich and rewarding area for teachers and students to explore. Rather than seeing science and social studies as subjects in which everything is known and all the issues are settled, students have an opportunity to examine scientific and societal questions over which disagreement exists. In fact, having students reflect on their own evolving definitions of the different conceptions of justice presented in this chapter is a worthwhile exercise in itself. For example, *Environmental Health perspectives* provides us with a sample lesson in which students discuss whether environmental health is a basic human right and, if so, what are its characteristics (see: http://www.ehponline.org/science-ed). Such an activity could be used regarding any of the different versions of justice described in this chapter.

As we will describe, individuals and groups hold conflicting definitions of social justice and those differences often underlie disagreements about what should be societal aims and how they should be achieved. These conflicting definitions also underlie differences between nations about what basic rights people should have and the role of government in providing and protecting those rights. For example, the nations in the European Union have generally taken a more proactive approach than the United States to limiting toxicants in the environment. As we noted in the previous chapter, the European Union has adopted the Precautionary Principle that states "when an activity raises the threat of harm to human health or to the environment, precautionary measures should be taken even if cause and effect relationships are not fully established scientifically" (Agyeman 2005, p. 20). Furthermore, European nations are much more likely to argue that justice requires that people have a right to a healthy environment and should do so without using more than their appropriate share of the world's resources (Agyeman 2005, pp. 104–105).

While the language of environmental and ecological justice emerged only in the last quarter century, environmental concerns have been present throughout recorded history. Hughes (1994), in *Pan's Travail*: *Environmental Problems of the Ancient Greeks and Romans,* shows how the Romans and Greeks were aware of and concerned about, among other environmental problems, airborne pollution. Similarly, throughout the last several centuries, many have raised concerns regarding urban environmental health (see, for example, Jacob Riis' *How the Other Half Lives*, Riis 1890/1966), and occupational health (for example, Upton Sinclair's *The Jungle*, 1946/1906).

However, a specific concern and language regarding human rights and particularly the right to good health emerged with the rise of socially democratic administrations after the Second World War (Judt 2005). In 1944, President Franklin Roosevelt proposed a Second Bill of Rights to extend the original Bill of Rights declaring that governments need to insure the following:

- The right to a useful and remunerative job;
- The right to earn enough to provide adequate food and clothing and recreation;
- The right of every businessman, large and small, to trade in an atmosphere of freedom from unfair competition and domination by monopolies at home or abroad;

- The right of every family to a decent home;
- The right to adequate medical care and the opportunity to achieve and enjoy good health;
- The right to adequate protection from the economic fears of old age, sickness, accident, and unemployment;
- The right to a good education (Sunstein 2004).

While Roosevelt's Second Bill of Rights is mostly forgotten, his widow, Eleanor Roosevelt chaired the United Nation's committee on Human Rights that included many of the same plus additional rights in the Declaration of Human Rights adopted by the General Assembly in December 1948 (http://www.un.org/Overview/rights.html). The Declaration of Human Rights was one of the inspirations for the more extensive charter of Fundamental Rights of the European Union, which the European Parliament has recommended be incorporated into the European Treaty (see the Human Rights Education Association at http://www.hrea.org).

Such efforts, concludes Brian Barry, in *Why Social Justice Matters* (2005), led to a general agreement after the Second World War that social justice included:

> (1) Ensuring that people came before profit, therefore making sure that workers could belong to strong trade unions and the public utilities and transportation be publicly owned to ensure that they were available to all; (2) Ensuring that income and wealth inequalities were not excessive and that a minimum level be provided for all; (3) Universally providing education and health services, thus eliminating the market criterion of 'ability to pay' (p. 6).

Barry argues for a conception of social justice that begins with reducing economic inequality, claiming that inequalities of wealth and income must be kept within a narrow range if equal prospects for education, health, and autonomy are to be realized. Barry's definition, like Roosevelt's, focuses on the individual's right to specific assets and resources such as a job, income, health and education.

However, the social democratic policies supported by Roosevelt and implemented in most industrialized countries have been greatly weakened by the rise of neoliberal policies over the last quarter century (see Harvey 2005; Hursh 2008). Neoliberal policies, initiated in the United States by Ronald Reagan, in the United Kingdom by Margaret Thatcher, and in Chile by Augusto Pinochet, aim to reduce the role of government in ensuring human welfare and to leave social decisions to the marketplace. Writers like Thomas Friedman, *in The World is Flat*: *A Short History of the Twenty-first Century* (2005), argue that deregulated markets provide the opportunity for everyone to become entrepreneurial and gain wealth, and, therefore, government should play a minor role in providing for human welfare. However, the current global economic crisis has led to some prominent proponents of neoliberalism, such as Alan Greenspan, long-time Chairman of the Federal Reserve and proponent of market deregulation, to admit that such policies have failed (Andrews 2008).

But, citizens across the globe are increasingly contesting neoliberalism and reasserting notions of justice beyond market-based decisions. For example, thousands of people, including union members, environmentalists, and human rights groups, protested the World Trade Organization at its meeting in Seattle in

1999 and have continued to protest at the World Economic Forum's annual meetings. More recently, the economic crisis brought on by investment firms and banks making unwise investments and loans have resulted in some calling for a reassertion of government regulation into the economic order. Some have characterized the political nexus as the struggle to balance freedom and order (Beinart 2008). Similarly, as we describe below, proponents of environmental justice have become more vocal and visible across the globe (Agyeman et al. 2005; Carruthers 2008), claiming their right to a healthy environment.

One attempt to develop a global perspective on the environment and health was the passage, in 2000, of The Earth Charter. As described on their website of the Earth Charter Initiative (http://www.earthcharterinaction.org):

> The Earth Charter is a declaration of fundamental ethical principles for building a just, sustainable and peaceful global society in the 21st century. It seeks to inspire in all people a new sense of global interdependence and shared responsibility for the well being of the whole human family, the greater community of life, and future generations. It is a vision of hope and a call to action.

The process of writing and agreeing on The Earth Charter began in 1987 and the final result is "a product of a decade-long, worldwide, cross cultural dialogue on common goals and shared values. The Earth Charter was finalized and then launched as a people's charter by The Earth Charter Commission (2000), an independent international entity" (The Earth Charter Initiative).

The Earth Charter, which is available in dozens of languages at the website listed above, has 16 principles grouped under four main themes: respect and care for the community of life, ecological integrity, social and economic justice, and democracy, nonviolence and peace. An example of one of the principles is: (Number 6) Prevent harm as the best methods of environmental protection and, when knowledge is limited, apply a precautionary approach. Under this principle are five sub-principles that include ensuring "that decision making addresses the cumulative, long-term, indirect, long distance, and global consequences of human activities," and "prevent pollution of any part of the environment and allow no build-up of radioactive, toxic, or other hazardous substances."

Like the other descriptions of environmental justice presented in this chapter, the Earth Charter could be analyzed in terms of how it should be interpreted and its practical implications. For example, what does it mean to "prevent pollution of any part of the environment," would be a worthwhile discussion.

The Earth Charter and other activities mentioned above demonstrate that there has been a long history of resisting urban environmental and occupational health. In the United States the environmental justice movement is often described as beginning with two events: Love Canal in 1979 and Warren Country in 1982. Internationally, the 1984 Bhopal disaster in India, in which Union Carbide leaked a lethal gas from its factory, killing more than 10,000 and disabling hundreds of thousand (Satyanand 2008) is one of the world's worst industrial accidents and, for which the legal and environmental justice debates continue (International Campaign for Justice in Bhopal: http://www.bhopal.net).

Scholsberg (2007) describes the community response at Love Canal as signifying the shift from seeing toxicants as a threat to individuals to a threat to which communities must respond. The events in Warren County, North Carolina, are credited with inciting the movement against environmental racism and for environmental justice. Schlosberg (2007) and Shrader-Frechette (2002) describe how the residents in Warren Country, one of the poorest countries in North Carolina and with a population that was 65 percent African American, successfully opposed the dumping of PCB-laden dirt in a new hazardous waste landfill.

Similarly, the poor in the global North and South have always faced environment health risks posed by inadequate water and sewage systems. Mike Davis, in *Planet of Slums* (2006), describes how more than half the world's population now lives in cities and the poor are concentrated in vast slums on the edge of those cities. The lack of clean water and proper waste disposal are responsible for most of the diseases. Davis quotes public-health expert Eileen Stillwaggon (1998) as stating, "Every day, around the world, illnesses related to water supply, waste disposal, and garbage kill 30,000 people and constitute 75 percent of the illnesses that afflict humanity (p. 95)." Davis adds, "Digestive-tract diseases arising from poor sanitation and the pollution of drinking water—including diarrhea, enteritis, colitis, typhoid, and paratyphoid fevers—are the leading cause of death in the world, affecting mainly infants and small children" (pp. 142–143). These diseases are preventable, if we only provided the world's poor with clean water and an effective way of disposing of their waste.

David Carruthers, whose research focuses on Latin America, writes in *Environmental Justice in Latin America: Problems, Promise, and Practice* (2008),

> Environmental injustice is real to the millions who breathe the poisoned air of Mexico City, Santiago, and Sao Paulo. It is real to the farm workers and day laborers who spend their days and nights in the pesticide drenched fields of Mexican, Central American, and Chilean export agriculture.... And it is real in the shadows of the export factories of Mexico, Honduras, and Nicaragua, where shantytown families store drinking water in discarded chemical barrels and raise their children in a landscape leached through with health metals (p. 7).

Carruthers (2008) raises several additional issues that will be considered in this chapter. First, he points out that those most likely to be negatively affected by environmental hazards are those who have been deliberatively excluded "from the political decisions that determine the locations and the risk levels of environmental threats" (p. 8). Second, the rise of global neoliberalism emphasizing free trade and reductions in governmental social spending have resulted in "worsening economic polarization, the erosion of basic economic security, the collapse of small farms and businesses, and insurmountable household debt" (p. 11).

Similarly, in the United States Bob Bullard (1993, 1994), who has been a leading critic of environmental racism, questions whether environmental laws, regulations, and policies are adequately protecting people or are being fairly applied across all segments of the population. As Bullard writes,

> Some individuals, groups, and communities receive less protection than others because of their geographic location, race, and economic status. Generally, environmental problems in suburban areas pose far fewer public health threats than do those in urban or rural areas. Moreover, low-income communities and communities of color bear a disproportionate burden of the nation's pollution problems (1994, p. 16).

For Bullard and others, environmental justice requires more than understanding the problem but also acting on it. It is not enough to hope that government will protect us from different environmental health hazards because, as Bullard reminds us, government is often the problem. In response, writes Bullard, activists have "targeted disparate enforcement, compliance, and policy formation as they affect environmental and public health decision making on a wide range of issues, from toxic waste to urban transportation" (1994, p. 17).

Julian Agyeman, in *Sustainable Communities and the Challenge of Environmental Justice* (2005), acknowledges that environmental justice is a "contested and problematized concept" (p. 20) and begins with the definition provided by the Commonwealth of Massachusetts in its "Environmental Justice Policy":

> Environmental justice is based on the principle that all people have a right to be protected from environmental pollution and to live in and enjoy a clean and healthful environment. Environmental Justice is the equal protection and meaningful involvement of people with respect to the development, implementation, and enforcement of environmental laws, regulations, and policies and the equitable distribution of environmental benefits (Commonwealth of Massachusetts 2002, pp. 2, 26).

Agyeman notes that the definition includes several crucial aspects of environmental justice. In particular, it includes procedural aspects ("meaningful involvement of all people"), substantial justice aspects ("right to live in and enjoy a clean and healthful environment"), and distributive aspects ("equitable distribution of environmental benefits") (p. 26), which are all concepts that we will build on in this chapter.

Agyeman's (2005) goal is to extend the definition of environmental justice by including environmental sustainability, or what he labels the Just Sustainability Paradigm (p. 5). In doing so, he wants to extend environmental sustainability beyond its emphasis on conservation of the natural environment to include race, class, justice and equity. He defines Just Sustainability as "the need to ensure a better quality of life for all, now and into the future, in a just an equitable manner, while living within the limits of supporting ecosystems" (2005, p. 6).

Others, such as Richard Hofricheter (2000), desire to combine environmental sustainability with environmental justice. He states, "Environmental justice is about social transformation directed toward meeting human needs and enhancing the quality of life—economic equality, health care, shelter, human rights, species preservation, and democracy—using resources sustainability" (p. 4).

More recently, academics and activists have widened the definition of environmental justice even further to include not only human welfare but also the welfare of diverse cultures and all living things. The Center for EcoJustice Education broadens the goal to "revitalizing those practices, relationships, policies and institutions that lead to strong democratic and sustainable communities recognizing the interconnections among biological, cultural and linguistic diversity as critical strengths" (Center for EcoJustice Education, p. 1). Rebecca Martusewicz, center founder, and her co-author, Jeff Edmundson, write that ecojustice requires that teachers "begin to recognize the limits of social justice approaches that do not address our interdependence as humans upon threatened natural systems" (n.d., p. 2).

Bowers' many publications (2005, 2008b) have played a central role in conceptualizing ecojustice and calling for school curricula to focus on ecojustice. Bowers is critical of those who suggest that reducing economic inequality is either sufficient or the primary step in attaining an environmentally just and sustainable world. He critiques proponents of distributional justice, such as Barry above in *Why Social Justice Matters* (2005), and proponents of critical educational theories such as Dewey (1916,1948/1920), and Freire (1973). Instead, he argues that our current economic and ecological problems can only be solved through a substantial critique of the rise of western consumer culture and our amnesia regarding other more community and place-based cultures. Bowers states:

> The aspects of ecojustice that should be the focus of educational reforms at both the university and public level are connected with the need to reduce the impact of the industrial/consumer dependent culture on everyday life while at the same time ensuring that people are not impoverished and limited in terms of equal opportunity; the five aspects of ecojustice that have special significance for educational reformers include the following: (1) eliminating the causes of eco-racism; (2) ending the North's exploitation and cultural colonization of the South (Third World cultures); (3) revitalizing the commons in order to achieve a healthier balance between market and non-market aspects of community life; (4) ensure that the prospects of future generations are not diminished by the hubris and ideology that drives the globalization of the West's industrial culture; (5) reducing the threat to what Vandana Shiva refers to as "earth democracy"—that is, the right of natural systems to reproduce themselves rather than to have their existence contingent upon the demands of humans; ecojustice provides the larger moral and conceptual framework for understanding how to achieve the goals of social justice (n.d.).

For Bowers (2008a), then, the emphasis on economic redistribution by philosophers such as Barry (2005), above, and critical theorists, such as Freire and Dewey, fails to consider that we also need to develop more inclusive models of decision making in which everyone participates and to go beyond recognizing to understanding and including other ways of understanding the world.

Lastly, David Harvey (2009), a geographer, attempts to integrate place-based knowledge, what we know regarding the unique features where we live or work, with equity issues. He asks the we "answer the questions of what kind of daily life, what kind of relation to nature, which social relations, what production processes, and what kinds of mental conceptions and technologies will be adequate to meet human wants, needs, and desires" (p. 247). In sum, developing a world that is socially and environmentally just requires that we analyze almost every aspect of the human condition: how we live, work, play, and learn.

In the first part of this chapter, then, we have provided an overview of the way in which conceptions of justice have changed over the last half-century. Initially, social democratic justice and human rights focused on distributional forms of equality in which individuals were to be provided with minimum economic and other welfare benefits. However, beginning in the late 1970s, environmental justice began to focus on how certain groups in the global North, such as people of color and indigenous groups but also poor whites, and in the global South, the poor and the marginalized, were more likely to be affected by environmental hazards. Later, in the 1980s, environmental and educational activists began to more broadly define justice to include the environment and other cultures. Agyeman (2005) and Hofrichter (2000) combine environmental justice and environmental sustainability with the goal of developing a society in which the health of the environment for this and future generations are also considered. Both are also concerned with the process through which justice is achieved, emphasizing that the procedure needs to be fair and transparent. Martusewiscz (2008) and Bowers (2008a), in their promotion of ecojustice, explicitly add the concern that we learn from past cultures and each other to develop "community-centered alternatives to the deskilled individual lifestyle that is increasingly dependent upon consumerism" (Bowers 2008b, p. 158). Moreover, adds Bowers, ecojustice includes "the need to undertake educational reforms that address our responsibility for leaving future generations within sustainable ecosystems, which all means recognizing the right of non-human forms of life to reproduce themselves in sustainable ways" (p. 158).

In the second half of the chapter we provide in-depth descriptions of lessons and units in which elementary and secondary students examined environmental health within the context of environmental justice. We describe how fifth graders (11–12 year olds) researched the dangers from pesticides and herbicides, explored what they can do to reduce their own exposure, and then developed different media presentations to educate students, teachers, parents, and community members. We also describe students in a secondary social studies class who researched the problems associated with pesticide use, lead poisoning, electronic waste, and hazardous materials transport and then developed public policies to reduce or eliminate the dangers. They also researched why some communities face greater health risks than other communities and worked to create social policies that will reduce those risks. Next, we turn to an overview of the ethics module at AMBIENT (2004), the program at the University of Miami. In this module, secondary students are educated regarding the key concepts in ethical decision making as they explore case studies examining toxicants in the air, water, soil, and food. In all of these lessons, students learned to use ethical reasoning protocols to create productive and civil discourse in urban high schools.

Students can go beyond examining specific toxicants in the environment to working to lessen both individual and community risks. Furthermore, teachers can incorporate into their classrooms the concepts of environmental health, environmental racism, and environmental and ecological justice. We present an ethical reasoning protocol that has been used in urban classrooms to create productive and civil discourse about environmental health issues; an instructional strategy that is critical to teachers feeling comfortable addressing ethical issues in the classroom. We also review a curriculum module as an example of how a local, historical issue can be the basis of study that goes beyond its original context. Students discover why Virginia Key in Miami was the traditional "Black beach."

Teaching Environmental Justice

Is Environmental Health a Basic Human Right?

Teachers might begin by having students define environmental justice. Children are often keenly interested in notions of justice because of their own desire to be treated fairly by adults, including teachers and parents. Therefore, if organized properly, students are likely to be easily engaged in discussions about justice. Teachers might begin with some of the descriptions offered here, such as discussing what should be included in a definition of human rights and begin with either Franklin Roosevelt's Second Bill of Rights or the United Nations Declaration of Human Rights.

The lesson "Is environmental health a basic right?" developed by Wendy Stephan and Lisa Pitman at the University of Miami Rosenstiel School of Marine and Atmospheric Science, NIEHS Marine and Freshwater Biomedical Sciences Center suggests that students being by reading "Is Environmental Health a Basic Human Right" (Taylor 2004) and then attempt to attain consensus around which of the following UN Declaration of Human Rights should be rights. They offer the American Heritage Dictionary definition of a right as "something that is due a person or governmental body by law, tradition, or nature." The UN Declaration of Rights includes:

- Right to life, liberty and security of person
- Freedom from unjust imprisonment
- Freedom to move freely within one's country
- Freedom to leave and return to one's country
- Right to have families
- Right to own property
- Right to the protection of law
- Right to privacy
- Freedom of thought, religion, speech, and assembly
- Right to work and be paid fairly
- Right to an education
- Right to a standard of living adequate for health and well-being
- Right to a clean environment (i.e. clean water, air, etc.)

In discussing which of these should be a right, students should try to specify what the right intends. For example, what is meant by "a right to an education"? Elementary? Secondary? University? Should a university education be free to anyone who meets entry requirements? What would be a standard of living adequate for health and well being? Students could also ask whether health care should be a human right. What does it mean to have a right to a clean environment? Potable water? What is meant by unsafe?

Moreover, in the discussion above we expanded environmental justice beyond human rights to include notions of environmental racism, environmentally sustainable justice, and ecojustice. Students could be asked, therefore, to weigh whether it is just to dispose of environmental waste in neighborhoods where people have little power to resist, neighborhoods in which the poor and people of color typically reside? If not, how should we dispose of environmental waste? Or, students could reflect on who benefits from material resources and who pays the cost for their production. For example, who consumes more of the world's resources and who pays the cost in terms of environmental degradation? In the film *The Story of Stuff* (available for free at: http://www.storyofstuff.com), Annie Leonard (2010) uses animation to show the consequences of consumption on resource depletion, waste disposal, and health.

Examining Toxicants in the Environment and Developing Public Policy and Citizenship Skills

Another way of approaching environmental justice is to engage students in researching the dangers from toxicants in their neighborhood and communities. Educators too often overlook the local as a means of coming to understand larger school issues. However, David Gruenewald, and Greg Smith (2007), in *Place-Based Education in the Global Age: Local Diversity* (2007), describe the benefits of examining the local as a means of helping children "understand the processes that underlie the health of natural and social systems essential to human welfare" (p. 16).

Examining the local has the further benefit of making it possible for students to affect local behaviors and policies. Social studies and other curricula often focus on policy making at the state, provincial, or national levels, levels at which most citizens, not to mention students, have little impact. Instead, it is at the local level where students can participate in local governance and policy making, and impact the behavior of fellow citizens.

At the University of Rochester, the project "My Environment, My Health, My Choices," used problem-based learning and case studies as a means of improving students' understanding of environmental health. The project coordinators, David Hursh, Camille Martina and Dina Markowitz, worked with middle and high school teachers (grades 5–12) to develop interdisciplinary units in which students learned about environmental health hazards and what they might do to eliminate or reduce their exposure risks. We focused on the specific environmental health risks in the students' neighborhood so that students could situate their environmental health issues within the local history and politics. For example, students in older urban areas examined the risk of lead poisoning from lead paint in older homes (Rochester, New York, has the highest rate of childhood lead poisoning in the United States), while students from suburban to rural areas researched the dangers of pesticides and herbicides.

In one suburban school, fifth-grade teachers focused on the real-life problem of toxic run-off into a stream near their school and the potential consequences for water pollution and human health. The teachers placed the problem of increasing toxicants in their environment within the history and geography of their suburban neighborhood. The students began by examining both drawings and photos of their neighborhood as it changed from a landscape inhabited by Native Americans, to an agricultural area settled by Europeans in the 1850s, and finally to a residential suburb after World War II. By looking at the historical changes, they realized that how humans lived on the land changed over time and affected the relationship between the land and human health.

The students also examined maps to understand that the stream near their school became larger as it ran toward a bay that was part of Lake Ontario, one of the Great Lakes in North America. This provided students with a visual demonstration of how toxicants that entered the creek near their school could flow into the lake where they might swim and toxicants might eventually return in their drinking water.

Lastly, students, in teams of three, chose which environmental health hazards to focus on, for example pesticides, herbicides, and one that they came up with (and their favorite) pet waste, and presented possible remedies to these dangers. The culminating assignment required the students to develop websites, digital presentations, videos, or brochures describing the specific hazards and possible remedies (a description of this and the following unit, with handouts and rubrics, is available at: http//:www2.envmed.rochester.edu/envmed/EHSC/outreach/curriculum.htm).

Switching from elementary to secondary school, a civics teacher, Michael Fantauzzo (Hursh et al. 2010) has been integrating environmental health into his 12th grade *Participation in Government* course, which is the culminating high school social studies course in New York State.

For Fantauzzo, environmental health provides a key to involving students in local policy making. Students often do not perceive civics as a subject that they need to know. But, this unit helps students understand that citizenship requires they identify and address community goals and problems. Furthermore, citizens can often have the greatest impact at the local rather than national or global levels. Lastly, to paraphrase John Dewey (*Democracy and Education*, 1916), schooling is not merely preparation for life, it is life, and, therefore, should be a place in which students not only learn about what they might do as adults but engage in democratic practices as children.

This unit responds to two common problems civics teachers' face. First, while many students, when asked to define citizenship in a democracy, use the right buzzwords, like freedom, liberty, and individual rights, few students have grappled with the complexity of these ideas, nor have they been asked to apply them to authentic situations. Second, many of the social issues that students learn about in school, and specifically in social studies, such as war, poverty, and inequality, are likely to overwhelm students and make them apathetic. However, teachers have the opportunity to help students understand that citizenship may begin not at the national or global levels, but at the local.

Students can respond to the environmental health issues examined in this unit by changing their own behaviors, and those in their family or school, or working to affect local policies. At minimum, students will come to appreciate how a small change in their own actions and attitudes can be part of a collective effort to overcome larger problems that appear to be out of the realm of individual control. Moreover, students can become informed and support local efforts to reduce environmental risks from various toxicants. For example, many local communities, including those in the Rochester, New York area, have instituted new residential regulations aimed at reducing dangers from lead poisoning, and pesticide and herbicide applications.

Class assignments change and evolve over time and this particular project is no exception; however, the core of the project has remained the same. The unit begins with a class brainstorming activity in which students list examples of environmental health issues and public health. After discussing this list first in small groups and then as a whole class, students define each concept and consider how these two ideas might align or contradict each other. The students in the class are next given five different letters, one to each student, each letter written by a fictitious author asking the local authorities to investigate a potential problem and provide some course of action. The topics point to problems with pesticide use, lead poisoning, electronic waste, state recycling bottle deposit programs, and hazardous materials transportation. The details of the problem have been made to fit the local environment and have been personalized with well-known landmarks and features. For example, the letter regarding pesticide use reads as follows:

To Whom It May Concern

As a resident for over 50 years, I have become increasingly concerned about pesticide use in agricultural and residential areas. Many of my neighbors now have their lawns "treated," while local farmers have increased their use of pesticides.

Recently, I noticed a foul smell emanating from my faucet along with an acrid flavor from my tap water. Concerned about our well water, my husband and I conducted a quick survey of our land, which is bordered on the southern side by a large cornfield. The eastern and western borders of our property are residential and both of our neighbors have their lawn tended by a local lawn-care company. There is a small dairy farm across the street from us and, much to our surprise, a puddle of standing water appears to have run off adjacent to our property and has the same smell as our tap water.

I have asked the town re-examine its regulations regarding pesticide use in light of possible risks to residents.

Sincerely,
Ivanna Kleenup

From this point, students identify the environmental problem and begin to investigate the nature of the problem and the effects on human health. Most of their research is conducted via the internet. To assist students in conducting their research, a list of reliable resources, including the environmental protection agency (EPA) and NIEHS are provided. Students are also given two handouts to help them assess and organize the information they collect.

One handout, "Source Evaluation Form," requires that they: (1) summarize the content of the source and how it will be used; (2) assess the credibility of the source, including when the source was written, authors and/or organization sponsor; and (3) whether and why they would recommend the source to a concerned citizen. In this way students begin to analyze information for objectivity and scientific accuracy.

A second handout, "Environmental Health Issue Research Guide," asks students to record what they learned from the source regarding: (1) the extent of the environmental risks on living organisms, including infants and children, adults, and the elderly; and (2) whether there are any regulations at the local, state, and federal levels that need to be followed when addressing this environmental issue.

After an extended period of guided research, students propose a course of action and follow through with that action by writing and debating laws and speeches, creating educational tools, or developing their own public service announcements using iMovies, which they then present to the class. Requiring students to present to the class gives the students an audience for their writing, making authorship more meaningful, and exposes the entire class to all five environmental health problems.

This integrated unit results in students realizing that citizenship requires more than voting. For example, Paul (all student names are pseudonyms), who researched the danger of lead poisoning, recognized that citizenship means getting "more involved with what is going on in the government and your neighborhood," so the you could prevent harmful activities that "you will end up regretting." Sara, in investigating the problem of electronic or e-waste, learned that this course differed from her other social studies courses, where she was taught "other people's opinions." In this course, she learned that "there is no one right answer and no one way to do things…. [In] public policy there are different policies that you can choose and…every person will have an opinion about which one is best…. I think I have a bigger capacity to be more of a citizen now that I know this and I need to be more aware, involved, to have an opinion, and learn."

For Davis and Goodman (2003), working with secondary students in Miami, Florida, the challenge of getting urban high school students to have rigorous, heated and yet substantive discussions about environmental health and justice issues was clear. How do you direct adolescent passion into thinking critically rather than yelling louder? Working with teachers in the AMBIENT project, they developed an instructional protocol for ethical reasoning to help students develop the concepts of veracity, transparency, responsibility and justice and use them to build knowledge about a situation in which reasonable people disagree. The protocol uses a tightly structured lesson plan in which students answer questions about an issue to determine what they believe and why they believe what they believe and not something else.

Students are given a scenario such as the following:

Veracity	What do you know about this situation? What do you believe to be true? Why do you believe it and not something else?
Transparency	What don't you know? What hasn't been asked? Is this the whole truth? What questions have not been answered?
Responsibility	Who is responsible? For what? What could be done? What are the possible (not necessarily desirable) alternatives?
Justice	What should be done? By whom? Why is this the best ethical decision?

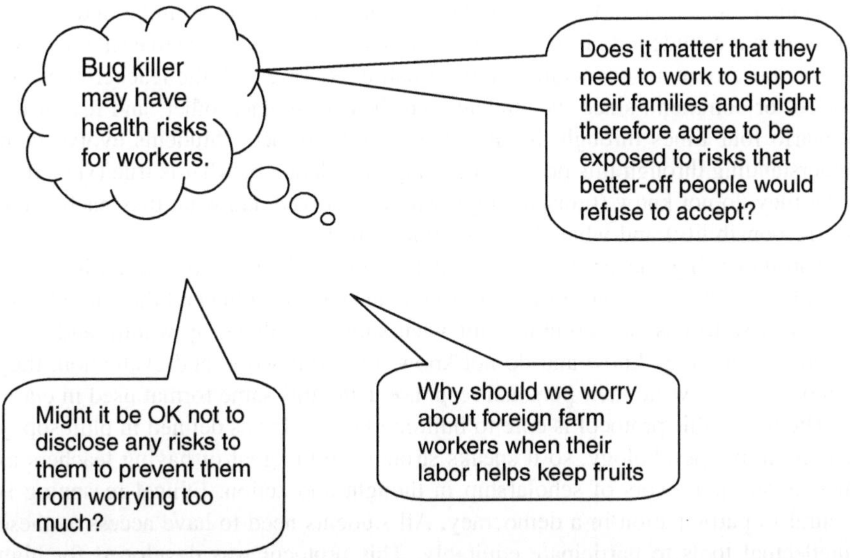

Fig. 1 Bug Killer Scenario (this and additional scenarios can be found on the website: http://www.rsmas.miami.edu/groups/niehs/ambient/teacher/ethics/ethics.html)

Bug Killer Toxicity

Some pollution is quite local. Methyl bromide is a gas used to kill bugs on tomatoes, strawberries and peppers. While it is injected into the soil, it also propagates through the air, sometimes spreading to nearby houses and communities. The chemical also damages the Earth's stratospheric ozone layer. The U.S. Environmental Protection Agency classifies methyl bromide as a "Class I acute toxin." In Florida and other states, the farm workers who come in contact with the gas are mostly poor people from Mexico, Central America and Haiti. Health risks range from nausea to headaches to cancer and memory loss … though there is not much research on the question (Fig. 1).

The students work in groups of three to come up with answers to each question, devoting about 7–10 min to each question. Each student in the small group has a role. One student, usually the most talkative, is asked to be the facilitator who makes sure everyone talks. A second student acts as recorder. The third student is tapped to be the reporter when each small group shares its ideas with the larger group. The initial discussion typically takes 45 min or what in the US is a regular class period. With a longer class period, the students can also do research to add to their evidence base. The goal is for students to use most of their time to either discuss the issues with one another or report out to the group.

The small groups report out to the whole, one after another without the teacher or other students commenting. They listen for ideas that are the same across all groups, and what is unique to only one group. Reporting out and listing common and unique ideas takes 4 or 5 min. The teacher records the common and unique ideas, or has a student do it, so everyone can see them in response to each question.

Even if erroneous or incomplete information is presented, the teacher does not correct or comment, since the students will hear from each other, and revisit the scenario four times through the questions in the protocol. Students evolve their understanding through this process, building knowledge of what is true (veracity), what they do not know (transparency), who is involved and what they each could do (responsibility) and what should be done (justice).

Through using the process, the students read and reread the scenario, listen intently to each other, and form their arguments so that in the end they are able to express a well-reasoned argument for what should be done, by whom, and why, based on what they know and do not know. For homework, and evaluation, they can be asked to write their personal response using this same format used in class.

The use of this protocol is core to building knowledge as defined in philosophy and cognitive psychology, so it speaks strongly to the goal of having teachers as models and purveyors of scholarship in thought and action. Ethical reasoning is central to participation in a democracy. All students need to have access to these intellectual tools to participate equitably. This protocol was developed for high school and undergraduate students from high poverty areas, and has been used with students in private and religious schools.

Students need a process that both pushes their thinking and directs their interaction to productive discourse. The ethical reasoning protocol presented here has been used with over 200 teachers and their students and the results are surprising in their similarity. Judgment and opinion are deferred and respect ensues as students read more carefully and listen more attentively to each other's ideas. Within the protocol they are living ethics because they are stakeholders, they want to be responsible or, at the very least, hold others responsible. They are interested in veracity and justice.

Through the use of the protocol students are immediately successful and interested in peeling back the layers of truth in the scenario through each round of discussion. They become interested in both the unique ideas that come up, and in the patterns that emerge. Being unique becomes a coveted prize so they push themselves to think out of the box, be funny or uncover some smoking gun. Recognizing the pattern of ideas, the ones that are repeated over and over again, gives them confidence that they are "getting at the evidence."

The interaction within and among groups turns out to be critical for keeping the attention of the group focused on the scenario. By using groups of three, it is difficult for students to hide, be quiet or drop out. By assigning them roles, they each have a part to play that they do not have to negotiate. By having the groups report out, with "ditto" and "pass" forbidden, they quickly hear what everyone else is thinking. By having them listen for unique and repeated ideas, they listen more carefully, comparing their group's work with the others. The pacing of the discussion and reporting rounds keeps students working against the clock to come up with something worth saying to the whole group. These are subtle elements, but critical to keeping students on task.

The use of a protocol for ethical reasoning is not unique. What may be unique about the protocol described here is the way it immerses students in the scenario and asks them to think in a very sophisticated ways without telling them that is what they are doing. Instead, they are doing what seems to come very naturally—digging for information and considering it from different angles, and in the process they engage in thoughtful and civil discourse about ethical issues.

Environmental Justice Module About Virginia Key

The issue of environmental justice on public beaches has a long history on Virginia Key in Miami. This beach ia located near where the overflow from a sewage treatment plant is sometimes discharged and has a vicious rip tides so swimmers are more likely to be swept away or drown than on other beaches. Moreover, not coincidently, the beach is a traditional "Black beach," left for African and Caribbean Americans.

In cooperation with the Audubon Society of Miami, the AMBIENT project created a module on the history, geography and tides of Virginia Key and the environmental health issues of drowning and sewage treatment. Over five full days, students conduct a beach clean-up to identify environmental health issues, develop a risk assessment, study satellite images and maps of the area, learn about drowning and sewage treatment, and then debate the question from different shareholder perspectives, "Should Virginia Key Beach be reopened to the public?" The activities can be used independently or as a module, on site or as a case study, or applied to another situation. The module is designed to have students learn firsthand about the risks associated with the tides of Virginia Key Beach to understand how environmental justice issues arise, the effects of unhealthy conditions, and what can be done. In the case of Virginia Key Beach, the prejudices of the dominant group and the need for access by an oppressed group conspired to create an unjust and dangerous situation for African Americans.

Students also conduct research on drowning to discover from the online National Center for Injury Prevention & Control (CDC) Drowning Fact Sheet (http://www.cdc.gov/HomeandRecreationalSafety/Water-Safety/waterinjuries-factsheet.htm) that African Americans and Alaskan Natives are more likely to drown than whites, especially children. Students learn the Environmental Protection Agency's definition of environmental justice and apply it to this situation:

> Environmental Justice is the fair treatment and meaningful involvement of all people regardless of race, color, national origin, culture, education, or income with respect to the development, implementation, and enforcement of environmental laws, regulations, and policies. Fair Treatment means that no group of people, including racial, ethnic, or socio-economic groups, should bear a disproportionate share of the negative environmental consequences resulting from industrial, municipal, and commercial operations or the execution of federal, state, local, and tribal environmental programs, and policies. Meaningful Involvement means that: (1) potentially affected community residents have an appropriate opportunity to participate in decisions about a proposed activity that will affect their environment and/or health; (2) the public's contribution can influence the regulatory agency's decision; (3) the concerns of all participants involved will be considered in the decision-making process; and (4) the decision-makers seek out and facilitate the involvement of those potentially affected (http://www.epa.gov/oecaerth/basics/ejbackground.html).

Students learn that economically disadvantaged people tend to be more affected by environmental contaminants because of where they live or work. They may be affected by proximity or exposure to waste sites, air pollution, pesticides, or wastewater contamination from city sewers or agricultural run-off. Students learn to use resources that identify known contaminants in an area such as the "enviromapper" by the EPA http://www.epa.gov/enviro/emef/ and http://www.scorecard.org. From these data and their own water sampling and beach debris analysis activities, and remote sensing data, they make maps that aid in their decision making about mitigation of hazards, and ultimately whether or not the beach should be reopened.

Students also learn that environmental justice issues can be political and contentious. They are presented with two articles presenting opposing viewpoints on environmental justice and guided to read each one critically, identifying the main ideas, evidence, loaded language and stakeholders in each article. In *Saving Our Backyard—Toxic Waste in Small Louisiana Town* (Kashef 1999), environmental justice is pitted against jobs in a small Louisiana town. Students learn that a survey conducted by the Deep South Center revealed a pattern of discrimination. "There was almost a perfect correlation between race and proximity to industrial sites," says the center's director, Beverly Wright, M.D." They learn that Greenpeace and Tulane University Environmental Law Clinic filed, in 1997, the first of several complaints with the EPA against the Shintech company that led to successfully preventing the company from building in the town of Convent.

In *The Environmental Racism Hoax*, students read an argument against environmental justice as a legitimate issue. Author David Friedman (1998) claims environmental justice was invented by groups seeking empowerment, and not based on facts,

> The agency claims that state and local policies deliberately cluster hazardous economic activities in politically powerless 'communities of color.' The reality is that the EPA, by exploiting every possible legal ambiguity, skillfully limiting debate, and ignoring even its own science, has enshrined some of the worst excesses of racialist rhetoric and environmental advocacy into federal law.... Properly analyzed, the data revealed that waste sites

> are just as likely to be located in white neighborhoods, or in areas where minorities moved only after permits were granted. Despite sensational charges of racial 'genocide' in industrial districts and ghastly 'cancer alleys,' health data don't show minorities being poisoned by toxic sites. 'Though activists have a hard time accepting it,' notes Brookings fellow Christopher H. Foreman, Jr., a self-described black liberal Democrat, 'racism simply doesn't appear to be a significant factor in our national environmental decision making.'

Students compare and contrast the viewpoints, engage in additional research on their own and even conduct a debate on whether or not environmental racism exists. Through hands-on experience and research, reading and interviewing, students learn about environmental justice, how to identify local issues and come to conclusions, and how to take action through recommendations for improvement.

Environmental justice and its relation to health provides opportunities for educators to involve students in real questions about what they can and should do to reduce their own and other's health risks. Furthermore, such explorations encourage teachers and students to situate their own experiences within the local community, or within questions regarding their nation or the globe. Teaching for environmental justice makes it possible for students to understand the relevance of history, politics, and economics to their own and other's health. Indeed, environmental justice provides an excellent opportunity for developing an interdisciplinary understanding of the world around us.

References

AMBIENT (2004). Social Justice. http://www.rsmas.miami.edu/groups/niehs/ambient/teacher/env_justice/Tej.htm

Agyeman, J. (2005). *Sustainable communities and the challenge of environmental justice*. New York: New York University Press.

Agyeman, J., Bullard, R. D., Evans, B. (Eds.). (2005). Just sustainabilities: Development in an unequal world. Cambridge, MA: MIT Press.

Andrews, E. (2008). Greenspan concedes error on regulation. *New York Times*. Accessed at http://www.nytimes.com/2008/10/24/business/economy/24panel.html?_r=1&scp=4&sq=Greenspan&st=Search. October 28

Barry, B. (2005). *Why social justice matters*. Cambridge, UK: Polity Press.

Beinart, P. (2008, November 24). The new liberal order. *Time Magazine, New York*, pp. 30–32

Beinart, P. (2008, November 24). The new liberal order. *Time Magazine*, pp. 30–32

Bowers, C. A. (n.d.). Ecojustice Dictionary. Available at: http://cabowers.net/dicterm/html/CAdict010.html

Bowers, C. A. (2005). Education for a sustainable future: Mediating between the commons and economic globalization. Available at http://www.cabowers.net/CAbookarticle.php

Bowers, C. A. (2008a). *Why the theories of Dewey and Friere cannot contribute to revitalizing the commons*. Available at http://www.centerforecojusticeducation.org

Bowers, C. A. (2008b). *Transitions: Educational reforms that promote ecological intelligence or the assumptions underlying modernity?* Manuscript available at: http://www.cabowers.net/CAPress.php

Bullard, R. D. (1993). Anatomy of environmental racism. In R. Hofrichter (Ed.), *Toxic struggles: The theory and practice of environmental justice* (pp. 25–35). Philadelphia: New Society Publishers.

Bullard, R. D. (Ed.). (1994). *Unequal protection: Environmental justice and communities of color*. San Francisco: Sierra Club Books.

Carruthers, D. V. (2008). Introduction: Popular environmentalism and social justice in Latin America. In D. V. Carruthers (Ed.), *Environmental justice in Latin America: Problems, promise, and practice*. Cambridge MA: MIT Press.

Center for Ecojustice Education. At http://www.ecojusticeeducation.org/index.php?option=com_frontpage&Itemid=1

Commonwealth of Massachusetts (2002). *Environmental justice policy*. Boston: State House.

Davis, H. & Goodman, K. (2003). *Ethics Module*. Miami: University of Miami Rosenstiel School of Marine and Atmospheric Science. http://www.rsmas.miami.edu/groups/niehs/ambient/teacher/ethics/ethics.html

Davis, M. (2006). *Planet of slums*. New York: Verso.

Dewey, J. (1916). *Democracy and education: An Introduction to the philosophy of education*. New York: Macmillan.

Dewey, J. (1948/1920). *Reconstruction in philosophy*. Boston: Beacon Press.

Freire, P. (1973). *Education for critical consciousness*. New York: Seabury Press.

Friedman, D. (1998). *The environmental racism hoax*. Washington, DC: The American Enterprise Institute.

Friedman, T. (2005). *The world is flat: A brief history of the twenty-first century*. New York: Farrar Straus & Giroux.

Gruenewald, D. A., & Smith, G. A. (Eds.). (2007). *Place-based education in the global age: Local diversity*. New York: Lawrence Erlbaum Associates.

Harvey, D. (2005). *A brief history of neoliberalism*. Oxford: Oxford University Press.

Harvey, D. (2009). *Cosmopolitanism and the geographies of freedom*. New York: Columbia University Press.

Hofricheter, R. (Ed.). (2000). *Reclaiming the environmental debate: The politics of health in a toxic culture*. Cambridge, MA: MIT press.

Hughes, J. D. (1994). *Pan's travail: Environmental problems of the ancient Greeks and Romans*. Baltimore: Johns Hopkins University Press.

Hursh, D. (2008). *High-stakes testing and the decline of teaching and learning: The real crisis in education*. New York: Rowman and Littlefield.

Hursh, D., Martina, C. A., & Fantauzzo, M. (2010). Toxins in our environment: health and civic responsibility. In E. Heilman (Ed.), *Social studies and diversity teacher education: What we do and why we do It*. New York: Routledge.

Judt, T. (2005). *Postwar: A history of Europe since 1845*. New York: Viking.

Leonard, A. (2010). *The story of stuff: How our obsession with stuff is trashing the planet, our communities, and our health—a vision for change*. New York: Free Press.

Kashef, Z. (1999). *Saving Our Backyard—toxic waste in small Louisiana town*. New York: Essence Communications New York.

Martusewiscz (2008). *Ecojustice in education*. http://www.centerforecojusticeducation.org

National Institute of Environmental Health Sciences (2005). From website: http://www.niehs.nih.gov/

Riis, J. (1890/1966). *How the other half lives: Studies among the tenements of New York*. New York: St. Martin's Press.

Satyanand, T. (2008). Aftermath of the Bhopal accident. *The Lancet, 371*(9628), 1900. June.

Schlosberg, D. (2007). *Defining environmental justice: Theories, movements, and nature*. Oxford: Oxford University Press.

Shrader-Frechette, K. (2002). *Environmental justice: Creating equality, reclaiming democracy*. Oxford: Oxford University Press.

Sinclair, U. (1946/1906). *The jungle*. New York: The Viking Press.

Stillwagon, E. (1998). *Stunted lives, stagnant economies: Poverty, disease, and underdevelopment*. New Brunswick, NJ: Rutgers University Press.

Sunstein, C. R. (2004). *The second Bill of Rights: FDR's unfinished revolution and why we need it more than ever*. New York: Basic Books.

Taylor, D. A. (2004). Is environmental health a basic human right? *Environmental Health Perspectives, 112*(17), A1003–A1005.

The Earth Charter Commission (2000). *The Earth Charter.* Accessed at: http://www.earthcharterinaction.org/content/pages/Read-the-Charter.html

Erratum to: Teaching Environmental Health to Children

David W. Hursh, Camille A. Martina, Michael A. Trush and Hilarie B. Davis

The first name of author Hilarie B. Davis was spelled incorrectly (as 'Hillary') throughout the original publication. The error occurs on the cover of the book and on the following pages: iii, iv, vii, ix, x, 39, 63 and 89.

D. W. Hursh (✉)
Warner Graduate School of Education, University of Rochester, Rochester, NY 14627-0425, USA
e-mail: dhursh@warner.rochester.edu

D. W. Hursh
Center for Globalization and Sustainable Development, The Earth Institute, Columbia University, New York, NY 1005, USA

C. A. Martina
Community and Preventive Medicine, School of Medicine and Dentistry, University of Rochester, Rochester, NY 14642, USA
e-mail: camille_martina@urmc.rochester.edu

M. A. Trush
Department of Environmental Health Science, Johns Hopkins Bloomberg School of Public Health, North Wolfe St. 615, Rm. E7537, Baltimore, MD 21205, USA
e-mail: mtrush@jhsph.edu

H. B. Davis
Technology for Learning Consortium, Inc., Sauga Ave. 75, North Kingstown, RI 02852-4515, USA
e-mail: hilarie@techforlearning.org

D. W. Hursh and C. A. Martina, *Teaching Environmental Health to Children*, SpringerBriefs in Environmental Science, DOI: 10.1007/978-94-007-1811-1_5,

Further Reading

Toxicology

Books

Eaton, D. L., & Gilbert, S. G. (2007). "*Principles in Toxicology.*" In C. D. Klassen (ed.), *Cassert and Doull's Toxicology: The Basic Sciences of Poisons.* New York: McGraw-Hill.

Organizations

Agncy for Toxic Substance and Disease Registry (ATSDR), http://www.atsdr.cdc.gov. ATSDR maintains data on hazardous chemicals at http://www.astdr.cdcgov/toxfaq.html and http:// www.atsdr.cdc.gov/toxpr2.html

Center for Alternatives to Animal Testing (CAAT) at Johns Hopkins University, USA. http://caat.jhsph.edu

Pesticides

Organizations

U.S. Environmental Protection Agency. "About Pesticides" http://www.epa.gov/pesticides/about/index.htm

World Health Organization

"*WHO commemorates 50 years of its Pesticide Evaluation Scheme (WHOPES), established with the approval of the World Health Assembly in 1960. The programme serves as a reference for setting norms and standards for public health pesticides and their life-cycle management.*"http://www.who.int/whopes/en/

European Environment Agency http://www.eea.europa.eu/

D. W. Hursh and C. A. Martina, *Teaching Environmental Health to Children*, SpringerBriefs in Environmental Science, DOI: 10.1007/978-94-007-1811-1,

Water

Books

World Health Organization. (1993). *Guidelines for Drinking Water Quality*, Vol.1: *Recommendations* (2nd ed.). Geneva: World Health Organization.

Organizations

World Health Organization. (WHO) http://www.who.int/water_sanitation_health/dwq/guidelines/en/

U.S. Environmental Protection Agency (EPA), http://water.epa.gov/drink/info/index.cfm

European Environment Agency http://www.eea.europa.eu/

United Nations

http://www.un.org/waterforlifedecade/

World Bank

http://web.worldbank.org/WBSITE/EXTERNAL/TOPICS/EXTWAT/0,contentMDK:21706928~menuPK:4602430~pagePK:148956~piPK:216618~theSitePK:4602123,00.html

U.S. Geological Survey (USGS)

"*The National Water-Quality Assessment Program (NAWQA) provides an understanding of water-quality conditions; whether conditions are getting better or worse over time; and how natural features and human activities affect those conditions*." http://water.usgs.gov/nawqa/

World Water Council

"*World Water Council: An International Multi-Stakeholder Platform for a Water Secure World established in 1996 in response to increasing concern from the global community about world water issues. Its mission is to promote awareness, build political commitment and trigger action on critical water issues at all levels, including the highest decision-making level, to facilitate the efficient management and use of water in all its dimensions and on an environmentally sustainable basis*." http://www.worldwatercouncil.org/

Air Pollution

European Environment Agency http://www.eea.europa.eu/ and http://www.eea.europa.eu/themes/air

U. S. Environmental Protection Agency, Transportation and Air Quality http://www.epa.gov/otaq/

The AIR Now Web site http://www.airnow.gov/

"*The U.S. EPA, NOAA, NPS, tribal, state, and local agencies developed the AIRNow Web site to provide the public with easy access to national air quality information. The Web site offers daily AQI forecasts as well as real-time AQI conditions for over 300 cities across the US, and provides links to more detailed State and local air quality Web sites. Air Quality Forecasts—Nationwide daily air quality forecasts provided by State and local Air Agencies for over 300 major U.S. cities. Air Quality Conditions—Nationwide and regional real-time ozone air quality maps covering 46 US States and parts of Canada. These maps are updated daily every.*

Lead

U.S. Centers for Disease Control (CDC)
Environmental Medicine Case Studies on Lead Toxicity
http://wonder.cdc.gov/wonder/prevguid/p0000017/p0000017.asp
Agency for Toxic Substance and Disease Registry (ATSDR), *Who is a risk?* www.atsdr.cdc.gov/csem/lead/pbwhoisat_risk2.html *and Lead (Pb) Toxicity: Key Concepts | ATSDR—Environmental Medicine and Environmental Health Education—CSEM* http://www.atsdr.cdc.gov/csem/lead/pbcover_page2.html

Children's Health

Books

Wigle, D. T. (2003). Child Health and the Environment. New York: Oxford University Press. Etzel, R. A., & Balk, S. J. (Eds.), *Handbook of Pediatric Environmental Health.* (2nd ed.). Elk Grove, Ill.: American Academy of Pediatrics.

Journal

Environmental Health Perspectives (EHP). [Homepage.] http://ehp.niehs.nih.gov.
This journal is produced by the National Institute of Environmental Health Sciences (NIEHS) and has a monthly section dedicated to children's environmental health topics.

Agencies and Organizations

U.S. Environmental Protection Agency (EPA), Office of Children's Health Protection. http://yosemite.epa.gov/ochp/ochpweb.nsf/content/homepage.htm
Canadian Partnership of Children's Health and the Environment. (2005). *Playing it safe: Childproofing for Environmental Health.* A PDF of the brochure is available at http://www.healthyenvironmentforkids.ca
Cooper, K. (2005). *Child Health and the Environment- A Primer. Canadian partnership of children's health and the environment.*

The primer aims to present information to parents, caregivers, health practitioners, and others who are concerned about the impact that the environment has on children's health. The primer provides an excellent overview of environmental health issues, including why children are more vulnerable than adults, the different biological systems (respiratory, endocrine, immune), the different vectors (air, food, water), and how to impact policy. A PDF of the primer can be downloaded at: http://www.healthyenvironmentforkids.ca/english/resources/card_file.shtml?x=2323.

Children's Environmental Health Network (CEHN), http://www.cehn.org

Children's Health Environment Coalition (CHEC), http://www.checnet.org.

The David Suzuki Foundation has made available three reports that provide international comparisons of air quality standards, pesticide regulations, and water quality standards. The reports (The Air We Breathe, The Food We Eat, and the Water We Drink), all published in 2006, are available as PDFs on line at http://www.davidsuzuki.org/health, or available for purchase.

Healthy Schools Network, Inc. (HSN), http://www.healthyschools.org.

Author Biographies

Hillary B. Davis Technology for Learning Consortium Inc, 75 Sauga AVe, North Kingstown, RI 02852, USA, e-mail: hilarie@techforlearning.org

Hillary B. Davis (Ed.D.) is the President and Senior Researcher at Technology for Learning Consortium Inc. She specializes in working with organizations and projects to develop, support and evaluate student-centered learning environments. Dr. Davis is the author of books on questioning, minds-on science, building a culture of literacy and digital literacy. She has led national evaluations for NASA, NOAA and NSF funded projects.

David W. Hursh Associate Professor, Warner Graduate School of Education, and Human Development, University of Rochester, Rochester, NY 14627, USA, e-mail: dhursh@Warner.Rochester.edu

David W. Hursh (Ph.D.) is an Associate Professor in the Warner Graduate School of Education and Human Development at the University of Rochester. His research and writing focus on education policy, neoliberalism, and teaching environmental sustainability and social studies. His most recent book is High Stakes Testing and the Decline of Teaching and Learning: The Real Crisis in Education (Roman & Littlefield). He has published in numerous journals, including American Educational Research Journal, the British Educational Research Journal, Educational Studies, and Educational Researcher. In the summer of 2010, he developed and taught lessons on energy and environmental sustainability in an elementary school in Kampala, Uganda.

Camille A. Martina Department of Community and Preventive Medicine, 601 Elmwood Avenue, Rochester, NY 14642, USA, e-mail: Camille_Martina@urmc.rochester.edu

D. W. Hursh and C. A. Martina, *Teaching Environmental Health to Children*, SpringerBriefs in Environmental Science, DOI: 10.1007/978-94-007-1811-1,

Camille A. Martina (Ph.D.) is a Research Assistant Professor in the School of Medicine & Dentistry at the University of Rochester. Dr. Martina's primary appointment is in Community and Preventive Medicine, Division of Social and Behavioral Medicine, with a secondary appointment in Environmental Medicine. She is also the Program Director of a National Institute of Health (NIH) Pathfinder grant, Researcher Resilience through Multidimensional Mentoring, in the Office for Faculty Development and Diversity at the University of Rochester. She is also the lead program evaluator for the National Center for Deaf Health Research (CDC) and the University of Rochester's Clinical Translational Science Institute (NIH). Her research interests include prenatal exposures of environmental endocrine disrupting compounds and behavioral outcomes, social determinants of academic success of urban, low income and underrepresented minority students and career advancement of underrepresented academic researchers as well as program evaluation.

Michael A. Trush Department of Environmental Health Sciences, Johns Hopkins Bloomberg School of Public Health, 615 North Wolfe St., Rm E7537, Baltimore, MD 21205, USA, e-mail: mtrush@jhsph.edu

Michael A. Trush (Ph.D.) is currently Professor of Toxicology and Director of Community Outreach and Education Core at the Johns Hopkins Center in Urban Environmental Health and Johns Hopkins Bloomberg School of Public Health. His current activities include researching the mechanisms of toxicology, teaching Public Health Toxicology to graduate students and an environmental health course to undergraduate, and working on environmental public health in the Baltimore. He is also Deputy Director and Director of the Community Outreach and Education Core of the Johns Hopkins Center in Urban Environmental Health, a National Institute of Environmental Health Sciences Center. He is Associate Editor of the journal Progress in Community Health Partnerships: Research, Education, and Action.

Index

D. W. Hursh and C. A. Martina, *Teaching Environmental Health to Children*, SpringerBriefs in Environmental Science, DOI: 10.1007/978-94-007-1811-1,

MIX
Papier aus verantwortungsvollen Quellen
Paper from responsible sources
FSC® C105338
FSC
www.fsc.org

If you have any concerns about our products,
you can contact us on
ProductSafety@springernature.com

In case Publisher is established outside the EU,
the EU authorized representative is:
Springer Nature Customer Service Center GmbH
Europaplatz 3, 69115 Heidelberg, Germany

Printed by Libri Plureos GmbH
in Hamburg, Germany